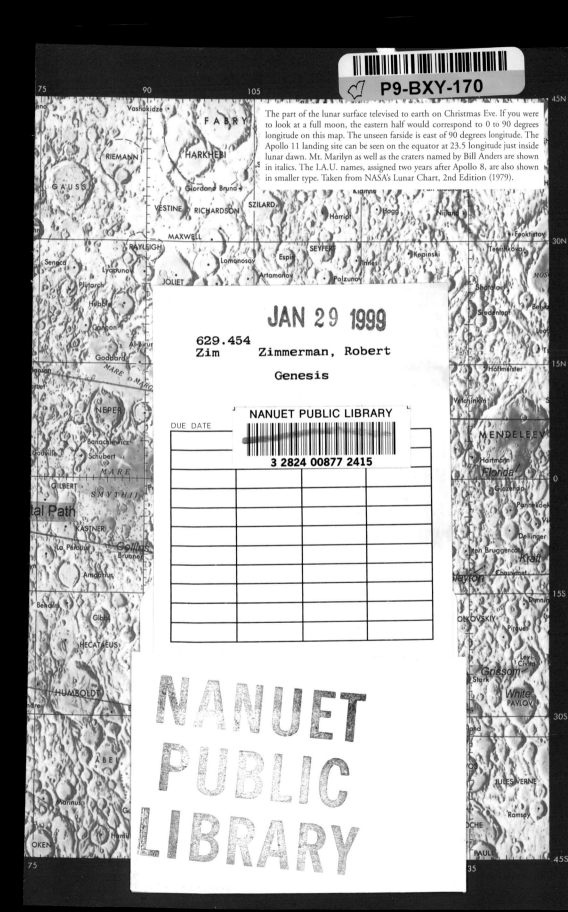

The part of the lunar surface televised to earth on Christmas Eve. If you were to look at a full moon, the eastern half would correspond to 0 to 90 degrees longitude on this map. The unseen farside is east of 90 degrees longitude. The Apollo 11 landing site can be seen on the equator at 23.5 longitude just inside lunar dawn. Mt. Marilyn as well as the craters named by Bill Anders are shown in italics. The I.A.U. names, assigned two years after Apollo 8, are also shown in smaller type. Taken from NASA's Lunar Chart, 2nd Edition (1979).

# GENESIS

## THE STORY OF APOLLO 8:
## THE FIRST MANNED FLIGHT
## TO ANOTHER WORLD

# GENESIS

## The Story of Apollo 8:
## the First Manned Flight
## To Another World

By Robert Zimmerman

Four Walls Eight Windows
New York/London

© 1998 Robert Zimmerman

Published in the United States by:
Four Walls Eight Windows
39 West 14th Street, room 503
New York, N.Y., 10011

U.K. offices:
Four Walls Eight Windows/Turnaround
Unit 3, Olympia Trading Estate
Coburg Road, Wood Green
London N22 6TZ, England

Visit our website at http://www.fourwallseightwindows.com

First printing September 1998.

Library of Congress Cataloging-in-Publication Data:
Zimmerman, Robert.
Genesis: The Story of Apollo 8: The First Manned Flight to Another World/ by Robert Zimmerman.
p.  cm.
Includes bibliographical references and index.
ISBN 1-56858-118-1
1. Project Apollo (U.S.) 2. Space flight to the moon.  I. Title.
TL789.8.U6A5285   1998
629.45'4'0973—dc21                                      98-29963
                                                        CIP

10 9 8 7 6 5 4 3 2

Text design by Acme Art, Inc.
Printed in Canada

**Front Cover:** This is Bill Anders's famous photograph of the first earthrise ever witnessed by humans, December 24, 1968, possibly one of the most reprinted lunar photos ever taken. If you turn the book sideways so that the moon's horizon is on the bottom, the picture will probably then look familiar, as this is how every publisher since 1968 has presented it. At Bill Anders's home, however, it is framed with the moon's horizon on the right. "That's how I took it," he says. To Anders, floating in zero gravity, the earth wasn't rising from behind an horizon line (which is how a human living on a planet's surface would perceive it). Instead, floating in a space capsule seventy miles above the moon, Anders saw himself circling the moon's equator. The lunar horizon therefore appeared vertical to him, and the earth moved right to left as it came out from behind the moon.

**Title Page:** The second earthrise ever witnessed by humans, framed by the command module window frame.

*To the memory of my uncle, Abraham Spinak,*
*who never stopped looking up.*

# SPECIAL ACKNOWLEDGEMENTS

Without the help, assistance, and support of the following people I could not have written this book: Frank and Susan Borman, Jim and Marilyn Lovell, Bill and Valerie Anders, my publisher John Oakes, Peter Brown, Steve Martin Cohen, Barbara Jaye Wilson, Peter Stathatos, Bob Moore, Jon Zimmerman and his family, and of course, my parents, who made both me and my book possible.

# TABLE OF CONTENTS

*Once a photograph of the Earth, taken from outside, is available, we shall, in an emotional sense acquire an additional dimension . . . Let the sheer isolation of the Earth become plain to every man whatever his nationality or creed, and a new idea as powerful as any in history will be let loose.*

Sir Frederick Hoyle, *astronomer, 1948*[1]

# INTRODUCTION

IN A SPACE BARELY LARGE ENOUGH FOR THE THREE MAN CREW, the astronaut opened the flight plan and began to read. "In the beginning," he said, "God created the heaven and earth . . ." Sweeping past him in the window was a stark black-and-white terrain, cold and forbidding. Unseen but listening intently was an audience of more than a billion people.

It was Christmas Eve, 1968. In what was probably the most profound Christmas prayer ever given by any member of the human race, Frank Borman, Jim Lovell, and William Anders read the first twelve lines of the Bible to a listening world as they orbited the moon, beaming homeward the first live television pictures of the lunar surface.

Though the flight of Apollo 8 was the first human journey beyond earth orbit and to another world, it has largely been forgotten in the ensuing decades amid the glory and excitement of the actual lunar landing in July of 1969. Yet this earlier mission probably exerted a much greater influence on human history. Not only did it signal the very first time human beings completely escaped the earth's gravity, but the astronauts used their bully pulpit in space to advocate the American vision of moral individuality, religious tolerance and mutual respect, while simultaneously giving us our first vision of the precious and unique lifegiving blue planet we live on.

To the people who participated in the 1960s space program, this flight is consistently referred to as the most important. Jim Lovell says, "I would rather have been on 8 than 11 . . . [Apollo 8] was the highpoint of my space career." Apollo astronaut Ken Mattingly told me "I consider Apollo 8 the most significant event . . . Compared to 8, [Apollo] 11 was anti-climactic."

And Neil Armstrong wrote that "Apollo 8 was the spirit of Apollo — leaving the shackles of earth and being able to return."[1]

For many like myself who watched from a distance, cheering these hardworking explorers on as they risked life and limb to reach another world, nothing compared with this first journey. As one magazine editor once told me, "Apollo 8 was my favorite mission." Anyone who followed the space program closely in the 1960's remembers the significance of Apollo 8.

This significance resonates in many ways. The power of television and the media was demonstrated more clearly than ever before. Almost two decades before the arrival of CNN, the television camera aboard Apollo 8 put nearly every person on the face of the earth in orbit around the moon. No longer was history made from a distance. Now every event would be seen, as it happened.

Apollo 8 also delineated the differences between the Soviet vision of society and the freely religious American system. Yuri Gagarin proclaimed he saw no god in space.[2] Borman, Lovell, and Anders saw Him everywhere, and said so. Whether one believes in God or not, the cynical desire of the now-collapsed Soviet empire to deny the spiritual magnificence of the universe could only have contributed to its fall. "There are more things in heaven and earth, Horatio,/Than are dreamt of in your philosophy."

The decision to read from the Bible was also not a governmental choice, but one that the astronauts, under Frank Borman's leadership, made entirely on their own. Their freedom to speak contrasted starkly with the Soviet Union and its state-run press and secret police.

The flight also demonstrated what today has become an almost blasphemous thought: that the peaceful competition between nations spurs achievement. We have forgotten the daring nature of this mission. It was the first (the very first!) use of the Saturn rocket to propel an object outside earth orbit. Yet, because NASA had heard of Soviet plans to fly two men around the moon before the end of 1968, the space agency decided to forgo any further tests and send humans to the moon as soon as possible.

Nor did this competition increase risks: neither nation could afford sloppiness if it wished to win the race. Though disasters happened, they did not occur often or with any greater frequency. Compared to today's leisurely but cooperative atmosphere, more was achieved successfully in a shorter time for less money.

This short weeklong journey to the moon also marked a crucial moment in both American and world history. In America, Apollo 8 put a positive, life-affirming exclamation point on what had been an ugly, violent year, with its political assassinations of Martin Luther King and Robert Kennedy as well as numerous urban riots and racial tension. In many ways, this mission signalled the actual end of the cultural sixties.

Worldwide, the experience of Borman, Lovell, and Anders illustrated for the first time how priceless the earth is to the human race. We have forgotten that before Apollo 8, no human had ever seen the earth as a globe. Now, suddenly, the human vision of the earth changed, and our mother planet became like all the others, a very small, lonely object in space.

Other cultural consequences, from a surge in environmentalism to the end of the Cold War, can be traced back to that single moment when Frank Borman, Jim Lovell, and Bill Anders read the opening words of the Bible.

For these reasons, I felt compelled to write this book. Hopefully I will not only succeed in giving these explorers their long overdue credit, but will also remind people of why we went to the moon in the first place. As Frank Borman said, "The Apollo program was just another battle in the Cold War." And, as writer Eric Hoffer noted, "It was done by ordinary Americans."[3]

# TIMELINE

1948-49:      Berlin Airlift.

1950:      Frank Borman graduates from West Point, marries Susan Bugbee.

1952:      Jim Lovell graduates from Annapolis, marries Marilyn Gerlach.

1955:      Anders graduates from Annapolis, marries Valerie Hoard.

1957:      Sputnik launched.

**1961**

    **April:**      Failed Big of Pigs invasion of Cuba; Gagarin becomes first human in space.

    **May:**      Shepard becomes first American in space; Kennedy proposes landing a man on the moon by 1970.

    **August:**      Berlin Wall built.

    **September:**      Borman and Lovell join astronaut corps.

**1962**

    **August:**      Soviet group flight of Vostoks 3 and 4; Peter Fechter killed trying to escape East Berlin.

    **October:**      Cuban missile crisis.

**1963**

    **October:**      Anders joins astronaut corps

    **November:**      President John Kennedy assassinated. Vice-President Lyndon Johnson becomes President.

**1964**

    **August:**      Tonkin Gulf Resolution passed. United States enters Vietnam War.

    **October:**      Three man Voskhod 1 mission; Khrushchev ousted.

**1965**

| | |
|---|---|
| **March:** | Leonov performs first spacewalk from Voskhod 2. |
| **December:** | Borman and Lovell fly two week Gemini 7 mission, rendezvousing with Gemini 6. |

**1967**

| | |
|---|---|
| **January:** | Apollo 1 launchpad fire. |
| **April:** | Soyuz 1 crash. |

**1968**

| | |
|---|---|
| **January:** | Tet Offensive in Vietnam. |
| **March:** | Martin Luther King murdered; Apollo 6 test launch. |
| **April:** | President Johnson withdraws as Presidential candidate; Columbia University protests; George Low first proposes sending Apollo 8 to the moon. |
| **June:** | Robert Kennedy murdered. |
| **August:** | Soviet invasion of Czechoslovakia; Democratic Party convention in Chicago; Apollo 8 lunar mission is given its first serious review and planning. |
| **October:** | Apollo 7 manned mission; Soyuz 2 manned mission. |
| **November:** | Nixon elected President; final decision made to send Apollo 8 to the moon. |
| **December:** | Apollo 8 launched Saturday, December 21st, for arrival in lunar orbit on Tuesday, December 24th, Christmas Eve. |

CHAPTER ONE

# "GET A PICTURE OF IT."

*Someone once pictured humanity as a race of islanders who have not yet learned the art of making ships. Out across the ocean we can see other islands about which we have wondered and speculated since the beginning of history. Now, after a million years, we have made our first primitive canoe; tomorrow we will watch it sail through the coral reef and vanish over the horizon.*

—Arthur C. Clarke, 1947[1]

SEALED WITHIN A CRAMPED CONE-SHAPED COMPARTMENT about the size of a customized van, three men waited silently. They were perched three hundred sixty-three feet in the air. Below them pulsed 1,500 tons of powerful fuel, about to ignite in a fiery explosion.

The date was Saturday, December 21st, 1968. The time was 6:51 AM (C.S.T.). The place was Cape Canaveral, Florida, at that time called Cape Kennedy.

Commander Frank Borman sat in the left couch. For the last twenty years he had dedicated himself to the defense of his country. When his government called for experienced test pilots to risk their lives flying into space, he had taken the call, becoming one of America's first sixteen astronauts. Now the forty-year-old sat poised and tense, his left hand gripping the abort handle. Though it was his responsibility to terminate the flight should something go wrong, he was determined this would be a perfect mission. He had devoted the last two years of his life in pursuit of that goal.

Jim Lovell, the command module pilot, sat in the middle couch. This was Lovell's third space mission, and he was at that moment the world's most experienced spaceman. Since childhood all he had ever wanted to do was build and fly rocket ships. Also forty years old, he had joined the military in order to fly the most advanced aircraft, and had become an astronaut to fly even more advanced rockets. Now, fired up with anticipation, his eyes fixed on the display panel of the spacecraft's computer as he keypunched data into the program.

Bill Anders, lunar module pilot, sat on the right. For him this flight was the adventure of a lifetime, flying the biggest and fastest flying machine ever built. Thirty-five years old, and the crew's only rookie, he had waited five long years for this first flight into space. Now he sat there and scanned the instrument panel and the spacecraft's systems, making sure everything was running properly.

Below them quivered the Saturn 5 rocket, tested only twice before, and never before used to put human beings in space.

These men had volunteered to be here. They had spent their lives trying to fly as fast and as high as a human being could go. And now they were about to go faster and higher than any human had ever attempted.

Apollo 8 was like no other space mission. Of the twenty-seven previous manned launches, both American and Soviet, none had ever ventured higher than 850 miles in altitude. Like the ancient sailors who had hugged the coasts, afraid to venture out into the vast uncharted ocean, astronauts and cosmonauts alike had remained close to Mother Earth, circling her again and again as they tested their ability to survive in the hard vacuum and utter cold of space.

After reaching orbit on December 21st, however, Apollo 8 would spend less than three hours circling the earth. If all systems checked out properly, the astronauts would re-ignite their third stage rocket and send their spacecraft hurtling away from the earth and out into the endless blackness of space.

Unlike past space missions, these astronauts would actually be going somewhere. They would be going to the moon.

* * *

For the past week the three men had spent considerable time taking care of personal business. The Sunday before, Jim Lovell had attended church, then drove out to nearby Edgewater, Florida where his mother Blanch lived with her sister and brother-in-law. In one week she would be seventy-three years old, and her family now gathered for a combined early Christmas and birthday dinner of ham and scalloped potatoes.[2]

When Jim Lovell was twelve, his father was killed in a car accident. Blanch Lovell had raised Jim herself, a single mom with an only son. Now her only son was about to travel to the moon, and he spent almost the entire birthday meal diagraming the mission and explaining the technical details of the flight to her. Jim Lovell very much wanted his mother to come with him, if not in body at least in soul.

Two nights later Marilyn Lovell arrived in Florida with their four children. She arranged to stay in a nearby beachfront cottage, where she and the children could swim in the ocean and relax while avoiding the crush of reporters. Just as Jim wanted his mother to be with him in spirit, he and Marilyn had decided that the entire family should share this first moon flight together.

Thursday night Jim and Marilyn took a car and drove to a place where they could see the gigantic Saturn 5 rocket, lit up against the night sky. As they sat there in the dark, he explained what the launch would be like, how the rocket would twist sideways to avoid the launch tower and how its flight would curve eastward as the rocket rose.

He then handed her a black and white photograph of the moon, taken by one of NASA's unmanned scout ships. She looked at it, puzzled. "I'm going

Lovell

to name that mountain for you," he explained, pointing at a triangularly shaped mountain on the edge of the Sea of Tranquility. Jim knew that he couldn't have gotten where he was without Marilyn's support, and by naming the peak after her she would share in his glory.[3]

\* \* \*

Bill and Valerie Anders said their goodbyes four days earlier. Since Bill had become an astronaut Valerie had seen as many live launches as possible, often driving long hours with her kids from Texas to Florida. She found the liftoffs exciting and exhilarating. If it had been possible she would gladly have flown into space herself.

This time, however, they both decided she should stay home. They had five young children, ranging in ages from four to eleven, and both

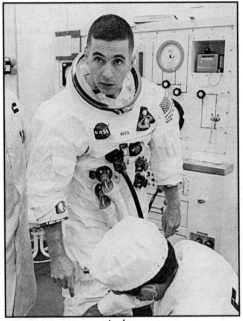

Anders

parents knew that a crush of reporters would descend upon them for this flight. And though they had just hired an au pair from Germany to watch their kids, Valerie could already see that eighteen-year-old Sylvie was not going to be much help. "All she did was clip news articles about the mission," Valerie remembered. It just seemed simpler to watch the launch from home.

Even so, she had flown out with Marilyn on Tuesday to spend a single evening with Bill. They had originally made their farewells just before Bill left for the Cape on December 8th. The night before they had had an early Christmas dinner, at which Bill had given Valerie and the kids their Christmas presents, including a color television to watch the mission. He also gave Valerie two audiotapes. One was a Christmas message he had recorded for the family that he wanted her to play on Christmas Day, while he was in lunar orbit. The other she was only to play if he did not return.

Then he gave her a farewell kiss and left for Florida.

One day later, President Johnson decided to have a party for the astronauts and their families, and Valerie and Bill unexpectedly found themselves together again at a formal dinner in the White House. Since it seemed ridiculous for them to say what might be their last goodbyes with so many people watching, Valerie made arrangements to fly out to the Cape to be with Bill one more time.

They spent that one short evening together. He went back to his crew quarters and she flew back to Houston and the kids.

Now it was less than twenty-four hours before launch, and Bill's childhood priest, Father Dennis Barry, came to visit him in his crew quarters. Barry was an old family friend, and Anders, a practicing Catholic, had invited him to the Cape to see the launch live. Bill showed him around, introducing him to the other astronauts. Then Barry performed a short private mass, giving Anders communion.[4]

Also visiting Bill that last day was Leno Pedrotti. Pedrotti had been Bill's thesis adviser when Anders had gotten his masters in nuclear engineering back in 1960. Anders had invited him and his brother Frank, a Jesuit priest, to come to Florida see the launch. The two brothers drove down from Ohio that Thursday to arrive at the astronaut's crew quarters just before sunset, and after introducing the two men to the other astronauts, Anders walked them to their car. As much as he wanted to, he couldn't spend much time with them. The launch was now only a few hours away.

They stood there in the parking lot, drinking soda and chatting about the mission and what Bill had done to get to this point. As they talked the sun was setting, and Anders noticed that a thin crescent moon floated slightly above it in the sky. The men shook hands and Bill was soon back in his quarters getting ready for bed.

It had only been a glance, yet Anders never forgot that crescent moon. The thought struck him that in mere days he would be there.

* * *

Frank Borman had his own intangibles to ponder. On Friday morning, just as the final countdown was beginning to gear up, Borman went to a local

Episcopalian church in nearby Cocoa Beach, Florida. There he spent a silent hour, finding solace and calm before the coming journey. Though he had the utmost confidence in his equipment and knew he had done all that was humanly possible to make sure everything would work, he felt a need to pray to God.

Then he went back to his room to call his wife one last time before the launch. Like Valerie Anders, Susan Borman had decided against seeing the launch live, and now waited at home in Houston with their two sons, Fred, seventeen, and Ed, fifteen.

They talked of the mission, and in his typical hardnosed test pilot manner, Frank reassured her that everything possible had been done to make the spacecraft and rocket flawless. He wouldn't be going if he didn't believe that.[5] Though Susan Borman immediately agreed with her husband, expressing confidence and encouragement for the man she loved, she actually did not believe him. For her, the risks surrounding the Apollo 8 mission seemed so obvious and incomprehensible that she no longer had faith in the success of the flight. She truly expected him to die in orbit around the moon.

She did not let Frank know of her conviction. He had chosen to risk his life for his country, and for eighteen years it had been her job to support him, regardless of the danger. As he had once said, and she had accepted, "There's more to life than just living."[6]

Frank did suspect that Susan was more frightened than usual. Christopher Kraft, Director of NASA Flight Operations, had come to visit her one night several weeks earlier, and they had talked about the flight and risks and dangers Frank faced. Normally, the astronaut wives never expressed their fears and doubts to anyone. For a wife to admit apprehension might lower her husband's standing in the eyes of NASA.

With Chris Kraft, however, Susan could be honest. He was a dear and close friend, and she trusted him. She also knew that he would tell her the truth.

She told him her fears. To her, it would be a miracle if NASA got them back alive.

Chris listened, and tried to give her some hope. He explained that the chances of success were probably fifty-fifty, not zero. He described the incredible effort that had gone into making the flight as safe as possible. He

Borman

talked of his faith in what he, Frank, and all of NASA were doing. Though Apollo 8 was clearly the most dangerous space mission ever, he was convinced it was the right thing to do.

For a couple of hours after Kraft's visit Susan considered what he had told her. *Maybe there is a chance,* she thought. *Maybe they will get home alive.* Then her doubts reasserted themselves. By launch eve she had dismissed what Chris had said and was once again resigned to despair.[7]

Aggravating her fears was the knowledge that she would be on public display, required to exude self-confidence and excitement not only for her husband but for the rest of the country. While she knew she could do it, it was a task and responsibility she found increasingly painful.

With the launch mere hours away, Susan lay in bed, unable to sleep. She waited for dawn and what she was convinced would be her husband's fiery launch into oblivion.

* * *

Frank also found sleep difficult that night. In his passionate and obsessive career his family had always come second to flying. Twelve years earlier, the Air Force had assigned him to teach at West Point, sending him first to the

California Institute of Technology in Pasadena to study aeronautical engineering. Borman wanted to fly planes, however, and he figured that the only way he could get that teaching assignment changed was to accumulate a lot more flight time. That Thanksgiving, he had the choice of staying home with his wife and two young sons or flying a total stranger to Washington, D.C. He hopped in his plane and went to Washington. For Borman, time in the air was far more important than time on the ground with his family.[8]

Nor was Susan resentful of this. She knew also that the success of their family depended on Frank flying as much as possible. Theirs was a lifetime partnership centered on the success of Frank's career.

To Frank, however, her unwavering support hadn't made the situation any easier. He loved his wife and children so deeply that if he thought of them while flying it distracted him, which in turn could be life-threatening. Hence, though he brought pictures of Susan and the boys with him into space, he refused to even glance at them, fearful of losing his focus.[9]

But Borman knew that his flying career could soar no higher, and weeks before he had told the space agency that Apollo 8 would be his last flight. In the future, he no longer wanted to do work that forced him to make believe his family did not exist.

Nonetheless, he still had to survive the flight of Apollo 8. As he lay there trying to sleep, he struggled to put his family from his mind, to focus his thoughts entirely on the needs of the mission.

* * *

As the sky began to lighten in the predawn hours of Saturday, December 21st, thousands and thousands of spectators gathered on the beaches of the Indian River. Some were NASA workers wishing to see the launch of their creation. Others were visitors from every part of the United States drawn by the thrill of this most audacious adventure. Many had driven to the beach the night before to sleep in their cars, guaranteeing themselves a good view.

Marilyn Lovell and her children had been picked up by a NASA official before dawn and driven to an isolated sand dune, only three miles from the launchpad. There, canvas chairs and food had been set up, and Marilyn and her children — Jeffrey, Susan, Jay, and Barbara, aged from two to fifteen — settled down to wait, bundled against the cool North Florida weather.[10] Accompanying them was Adeline Hammack, Marilyn's neighbor in Houston and one of her closest friends. Adeline had come with her to Florida to help keep watch on the children during launch.

Marilyn looked with awe at the giant rocket, amazed that at that moment her husband was perched in the tiny capsule at its peak. She knew how excited and happy he was, finally doing what he had always wanted to do. Though she had some apprehension about the risks, she did as he did, putting these fears aside to savor the thrill of man's first journey to the moon. Unlike Sue Borman, Marilyn had a deep abiding faith that everything would work properly and Jim would come home.

Also on the sand dune sat Charles and Anne Lindbergh. Two days earlier they had visited the astronauts for a pleasant afternoon lunch. The astronauts, about to join the ranks of pioneers like Lindbergh, sat in delight as he described his flight across the ocean. Later he talked about the time he had met Robert Goddard, the inventor of the liquid fueled rocket and the father of American rocketry. Everyone laughed when Lindbergh described how Goddard thought he could have designed a rocket to reach the moon, though "unfortunately it might cost as much as a million dollars."[11] The 1960s American space program, whose sole goal was to send a man to the moon, cost approximately twenty-five thousand times more.

Lindbergh himself was filled with wonder at what the astronauts were doing. At one point in the evening he calculated that in the first second of the Saturn 5's flight, it would burn "ten times more fuel than I did all the way to Paris."[12]

Several months later, Anne Lindbergh wrote about this short afternoon meeting with the astronauts. "Here in the midst of a scientific, mechanical, computerized beehive is the human element, the most exacting of all. There is nothing mechanical or robotlike about these men. Intelligent, courageous and able, they inspire faith in human capabilities and in this particular human exploit."[13]

All told, approximately a quarter of a million people had gathered to see the launch of Apollo 8. Standing on the shorelines surrounding Cape

Kennedy, they stared with equal wonder and dread at that huge missile sitting more than three miles away.

A thousand miles away, in the tiny suburban development of El Lago, Susan Borman sat on the floor of her living room, nibbling on sweet rolls, staring tensely at the television image of the Saturn 5 rocket. On the couch behind her sat her two teenage sons along with the family dog, Teddy. Numb and fearful, expectant of disaster and death, she waited silently for the mission to start, hiding her thoughts from all around her.

In that same El Lago development watched Valerie Anders with her five children. On their lawn stood a four foot high American flag. Fifty lights formed the stars, and other lights defined the red and white stripes. It had been designed and built by their neighbors, all NASA fellow workers, who the night before had stood on her lawn to say the Pledge of Allegiance, sing Christmas carols and the "Star Spangled Banner," and wish Bill and Valerie good luck.[14]

She and her children had begun their day when their priest, Father Vermillion, led them in a short private mass in their living room, with eleven-year-old Alan and ten-year-old Glen acting as altar boys in their stocking feet.

Father Vermillion ran the tiny Catholic chapel at Ellington Air Force Base to which the Anders family belonged. Because the Catholic community in Houston was so small, he had asked Valerie if her two oldest boys would serve as altar boys. From Valerie's point of view, Father Vermillion had to be desperate to enlist her children. At his first service, Alan held the chalice upside down so that the hosts dropped out, one by one, leaving a trail behind him like Hansel and Gretel. "Father Vermillion was a real laid-back man," Valerie remembered. "This just cracked him up."

Often he came by her home after Sunday services for lunch. Now, in the early dawn hours of launch day, he arrived as a friend, and gave the family communion. Valerie was glad he did so, particularly for the five young children. She knew they held deep unstated fears about what their father was doing, and felt that the priest's prayers would provide them comfort, telling them that the concerns of a wider world stood with them.

For her, the flight's excitement alone helped her bury her fears. Because she believed in what her husband wanted to do, thought it right and

Valerie Anders watches the countdown with Greg, Glen (on floor),
Gayle, and Eric.

Credit: Anders

worthwhile, she accepted the risks. Even when he told her that he only had a fifty-fifty chance of getting back alive, she refused to dwell on the possibility of his death in order to give him her fullest support.

She now sat with neighbors, friends, and other astronauts and talked with zest about the technical details of the mission and what was happening at each second of the countdown. Like the hundreds of thousands at the Cape, she sat in her living room and stared with wonder and amazement at the towering rocket, breathing white fumes and poised for takeoff.

It is difficult to understand today how truly gigantic the Saturn 5 was. Standing 363 feet high when fully assembled, this leviathan was twice as tall as today's space shuttle. If laid horizontally in any American football stadium it wouldn't have fit. The five engines at the base of its first stage generated more than seventy-five times the power of a 747 at takeoff.[15] Imagine the base

Credit: Anders

Alan Anders waits for lift-off.

of a forty-story skyscraper suddenly roaring into flames, lifting the building into the air and far into space.

In its first launch thirteen months earlier, the rocket's vibrational force sent tremors through the ground to the public viewing area — four miles away. The sound waves were so strong that television news anchor Walter Cronkite was forced to hold his television booth together with his bare hands to prevent the front plate glass window from falling on top of him.[16]

It was now just seconds before 6:51 AM, and in the tiny capsule on top of that missile Frank Borman, Jim Lovell, and Bill Anders tensed, waiting for the rumble of the rocket's engines and the pressure against their backs as the Saturn 5 roared skyward.

Nine seconds before launch, as intended, the five F1 engines of the Saturn 5's first stage ignited, sending gigantic clouds of smoke billowing sideways from the launch tower. For nine seconds clamps held the behemoth rocket in place, letting its engines build up thrust. Then, exactly as scheduled, the clamps let go, and the almost six million-pound Saturn 5 lifted free of the earth.

The Saturn 5 rocket, fully assembled and moving at 0.8 miles per hour
from the Vehicle Assembly Building to the launchpad. For scale, note man.

As the rocket rose, rookie astronaut Anders was startled how violent
and noisy the experience seemed, especially when compared to the simulators.
He was jerked from side to side as the Saturn 5's engines far below kept
adjusting the direction of their thrust to keep the rocket vertically balanced.[17]
This in turn whipped the rocket's tip, where the astronauts sat, back and forth
like an antenna.

To Lovell and Borman, however, the ride seemed almost comfortable
compared to past experience. Lovell thought it was unbelievably noisy. Borman

Marilyn Lovell, holding son Jeffrey, watches the Saturn 5 lift-off with daughters Susan, right, and Barbara, left. Son Jay is standing just off camera to the left.

later wrote that "the ride was incredibly smooth; compared to Apollo 8, Gemini 7 [his previous space flight] had traveled over a few hundred potholes."[18]

For just over two minutes the five F1 engines of the first stage pounded them upward, quickly accelerating them to a speed of over 6,000 miles per hour and an altitude of forty miles. Then, as planned, these engines cut off, the first stage was blown free, and after a moment's pause, the five J2 engines of the second stage kicked in. A moment later the rocket's escape tower and capsule shroud blew off. If they wanted, the astronauts could now look out their windows.

Not that they were really interested. When the first stage had cut off, all three men had been flung hard forward in their harnesses, then thrown backward violently when the second stage ignited. Anders thought he would fly straight into the instrument panel.[19] Borman worried about the strain on the rocket boosters. Lovell was merely exhilarated.

For six more minutes the second stage's smaller engines hammered them into the sky. Without the weight of the first stage to hold it down, the Saturn 5 accelerated much more quickly, its speed rising to more than 15,000 miles per hour. Then, at an altitude of over one hundred miles, these engines cut off and the second stage blew away, joining the first in its long fall down to the Atlantic Ocean.

Now the third stage ignited. For about two and a half minutes it burned, driving the spacecraft ever higher and faster. Only eleven and a half minutes after liftoff the third stage cut off, and exactly as planned, Apollo 8 was in a 115-mile orbit, moving at 17,400 miles per hour.

Other space missions would have now shed the third stage, becoming a space capsule just large enough to hold crew and supplies. In order to escape earth orbit, however, Apollo 8 needed its third stage. When refired, its fuel and engines would increase the spacecraft's speed by another 7,000 miles per hour.

In fact, that third stage could send the astronauts and their little space capsule as far as anyone wanted them to go. If NASA had wished it, those engines could have transported the Apollo capsule to any point in the solar system, from the moon to Pluto, and even beyond.

This time, however, the spacemen were merely going to the moon. For now, that would suffice.

* * *

For the next two orbits, both ground stations and astronauts labored hard to confirm that all systems checked out. At mission control, astronaut Mike Collins sat at capsule communications (or capcom for short), relaying information to and from the capsule. Collins had originally been assigned to this mission, and had been training for it since 1966. Then, in 1968, he began noticing that his legs no longer worked quite right. He found that his left knee sometimes buckled under him for no reason. He experienced a strange tingling and numbness in his left calf. The leg felt hot and cold at the strangest times. Very quickly he realized that these sensations were spreading.

Tests revealed a bonespur pressing against his spinal cord. Collins was immediately grounded, scrubbed from Apollo 8 and replaced by his backup Jim Lovell. And even though Collins's July operation was a complete success and by December his back had healed, he was still stranded on the ground. To his intense frustration, Collins found himself designated as one of the three astronauts assigned to handle the ground-to-capsule communications.[20] He sat there, both enthralled and envious of the men in orbit.

Cliff Charlesworth, the mission's prime flight director and the man whose job was to make all final decisions while on duty, stood nearby. Charlesworth, thirty-seven, was a quiet laconic Mississippian, known for his cool head and calm, deliberate manner. He now "called the roll," mission control jargon for going down the line of consoles and asking each man if everything in his department was "go," or "no go." One by one the men said "Go." Charlesworth turned to Collins and told him to let the astronauts know.

In a calm voice that contradicted the significance of his words, Collins announced that the astronauts now had permission to leave the earth. "Apollo 8, you are go for T.L.I." T.L.I. stood for trans-lunar injection, the five minute rocket blast that would send them toward the moon, accelerating their spacecraft to a speed of over 24,000 miles an hour, 7,000 miles an hour faster than any human had ever traveled.

Five seconds before ignition their primitive on-board computer (less powerful than the least sophisticated hand calculator available today) asked the astronauts one more time if they wished to abort. Jim Lovell calmly keyed in the command for the computer to proceed. At 9:41 AM the engines fired, burning more than eighty tons of fuel for five minutes and seventeen seconds. As their speed increased, Lovell called out Apollo 8's increasing velocity in feet per second: "Coming up on 28,000 . . . . 30,000 . . . 34,000 . . . 35,000 . . ."

Then the engines cut off, and they were flying away from the earth at 35,452 feet per second, or 24,171 miles an hour. If an ordinary jet airliner took off at that speed it would reach normal cruising altitude in one second.

At mission control Mike Collins told them, "We have a whole room full of people that say you look good." Chris Kraft couldn't resist exclaiming into the radio, "You're on your way — you're *really* on your way!"

They were no longer in earth orbit. Apollo 8 had become the first manned vehicle to break the bonds of earth.

\* \* \*

For the next half hour the three astronauts had little time to consider the magnitude of their journey. The third stage would automatically separate from the capsule in less than twenty-five minutes, and they had to be ready.

Borman and Lovell urgently scanned the many dials on the instrument panel, reading off the numbers for Anders to check against the flight plan.

Then, with what Frank Borman described as "a bone-jarring shock,"[21] the third stage blew free. For ten minutes Borman struggled to maneuver the command module so that the astronauts could spot the stage in their windows. Finally, he swung the ship around and Jim Lovell said, "There she is . . ." only to have his voice trail off in awe.

The third stage was clearly visible, barely five hundred feet behind them. Behind it, however, loomed the entire sphere of the earth. Their speed was so great that only forty minutes after leaving earth orbit, the planet had shrunk enough to be entirely visible within a single window.

All three men stared in silence. Even as they watched, the earth visibly shrank, the black velvet of space growing around it. To Lovell, the experience felt like he was driving a car into a dark tunnel: looking back he could see the light at the opening dwindle to a small speck behind him. Borman thought "this must be what God sees."[22] Anders was surprised at how delicate and pretty the earth looked.

After a minute of staring at the earth, Borman shook himself awake and decided someone should let the ground know. "We see the earth now, almost as a disk."

On the earth, Mike Collins had no idea how powerful that image looked to the astronauts. "Good show," he said. "Get a picture of it."

Lovell attempted to describe what they saw, outlining how he could see Florida, Africa, Gibraltar, and even most of South America. In one glance he could see the entire Atlantic Ocean.

Collins still hadn't quite sensed what the astronauts were seeing. He asked Lovell what window he was using, and again urged them to take pictures.

For thirty more seconds the three astronauts stared silently at the earth. Whatever their expectations for this journey, they had not anticipated this kind of vision. The earth's atmosphere lent it a translucent quality, almost as if it were glowing. And even as they watched, they could see it shrink in the surrounding darkness. In the ten minutes since Jim Lovell had first spotted it, they had moved over 3,500 miles farther away.

In less than three days, these men would reach lunar orbit. For a brief twenty hours they would circle the moon ten times, hold two press conferences, and generally try to relate to the awed world behind them their impressions of the first human exploration of another planet. And they would do this on Christmas Eve, the most significant spiritual holiday for almost a third of the world's population.

What would they say?

Surprisingly, the three astronauts had already chosen the theme for their most important message. Borman had found the words, and the others had completely agreed. All three knew where they stood in the cultural and political war that had been ongoing since the end of World War II and had become especially violent in the last twelve months. All three wished to contribute their thoughts on the matter.

Nonetheless, the incredible experience itself — the vastness of space, the desolation of the moon, and the lovely blue-white lure of earth shrinking steadily behind them — exerted its own inexorable command. Before these three men returned to earth, their experience and the words they spoke would have an influence on the world far beyond anything any of them had expected, or possibly even wanted.

CHAPTER TWO

# "WE WILL BURY YOU!"

## BORMAN

THE OVENS WERE STILL THERE. So were the gas chambers, the fences, the towers, the bleak dormitories. The metal sign on the gate still said "Work will set you free" in German. In the center of the Dachau concentration camp the gallows still stood, silent witness to the death of hundreds of thousands.[1]

The occupants, however, were no longer starving Jews condemned to mass slaughter. The year was 1949, and the refugees were from East Germany. Since the liberation of the concentration camp at the end of World War II, Dachau had been used first to shelter German prisoners of war and Soviet army deserters. In 1946, these deserters rioted. Ten actually killed themselves rather than be repatriated to the Soviet Union as per the Yalta accords.[2] Since then the camp had housed tens of thousands of German refugees who were fleeing the Soviet zones of occupation.[3]

Wandering through this oppressive place were a dozen West Point cadets, smartly dressed in uniform and led by Colonel Herman Bukema, head of the Academy's social science department. Bukema was giving his young charges a month-long tour of postwar Europe. They had already visited Cologne, seeing a city so flattened by bombs that it resembled the well-known pictures of Hiroshima. Soon they would visit Berlin, then Austria, Rome, Greece, and a dozen other places. For their transportation and living quarters Bukema had arranged the use of one of Hitler's private railroad cars.

Among Bukema's students was a twenty-one-year-old third year Cadet Corporal by the name of Frank Borman. Borman, raised in Tucson, Arizona, had never imagined that such deprivation was possible. The young man stared with dismay at the refugees. Their clothes were ragged and thin, and they had a beaten, tired look about them. Whole families were crowded into the dormitories, using blankets to cordon off their meager living quarters.

Nor were these the only horrors that he had seen. Beaten, occupied, and nothing more than shattered plunder for other nations to fight over, the citizens of Germany in the late 1940's could barely find enough food to eat. Cities lay in rubble from Allied bombing, and the lack of food had been worsened by a severe drought in 1947. Compounding these problems was the ceaseless tug-of-war between the Soviet Union and the other allies for control of Berlin and the reconstruction of Germany.

To this destitute land came Frank Borman, a blond, small-boned man whose friendly face belied his intense, dedicated and relentless mind. Born in Gary, Indiana in 1928, he had been a sickly child, with serious sinus problems. When their family doctor told his parents that their son had to leave the industrial Midwest for his health, his father gave up his successful auto repair shop and, in the worst years of the depression, moved his family to Tucson, Arizona. Unable to make a profit with a new gas station, Edwin "Rusty" Borman was forced to take odd jobs changing tires at someone else's garage, while Marjorie Borman rented out rooms in their home.

For Frank, however, these problems didn't exist. His childhood in the warm desert country of the American Southwest was like being in heaven. He wandered the countryside, bringing home strange pets, from goats to tarantulas. He and his father built homemade model airplanes, some with wingspans as long as six feet, and each Sunday morning they took the planes out to the wide open windswept desert fields and flew them far and high.

Flying was an early obsession for the boy. When Frank was five, before his parents moved to Tucson, his father paid a barnstorming pilot five dollars so that he and Frank could ride in an old biplane. The boy sat in the front cockpit with his father, feeling the wind in his hair and the unbounded freedom of the open sky and far horizon.

As Frank grew so did his love for flying. At fifteen he decided flying model airplanes was no longer sufficient: he wanted to fly himself. Though his parents didn't object, they insisted he pay the expenses on his own. Working three different partime jobs while attending high school, Frank earned enough each week to pay for one two-hour flying lesson each Saturday. Soon he was flying solo, having earned his pilot's license while still a teenager.

One Saturday Frank was caught in a sudden thunderstorm as he was returning from one of his first solo flights. Fighting the howling wind and the turbulence, Borman suddenly felt great excitement and joy: he was going to bring that plane home no matter what. His mind cleared, his senses became sharp, and he focused his entire being on doing what had to be done to land safely. When that plane glided to a stop at the end of the runway, Borman found himself overwhelmed with an extraordinary feeling of accomplishment.

By his senior year of high school Frank Borman knew that he wanted to spend his life flying airplanes. He had also met and dated the one woman he would share that life with.

Unfortunately, he didn't know this yet.

She did, however.

When seventeen-year-old Frank Borman first asked fifteen-year-old Susan Bugbee for a date, she knew that he was the man for her. Her father had died when she was thirteen, and she saw in Borman a stability and strength that few other teenagers had. She knew that he would be successful in whatever he did, and she fervently wanted to help him get there. Borman himself was strongly attracted to Susan. She was smart, articulate, and beautiful. By the end of high school they were going steady. Neither, however, had yet considered marriage.

Frank's focus was instead on flying. Because Borman's family was too poor to send him to any of the preeminent aeronautical schools, he was left with two choices: enlist and take advantage of the G.I. Bill, or apply to West Point. Unfortunately, he hadn't thought of West Point until well into his senior year of high school. It was now too late. There were too many applicants ahead of him.

Then, as Borman believes, fate intervened. The son of a local judge was in trouble, hanging out with the wrong people. Having heard that a certain high school student by the name of Borman had not only obtained his pilot's license but also built and flew model airplanes, the judge asked Frank to work with his son. He even offered to buy all the model plane kits, regardless of cost.

For Borman this was a deal he couldn't pass up. He and the boy became good friends as they assembled and flew some of the most expensive model planes available. In gratitude for straightening his son out, the judge pulled the right strings and got Borman on the applicants' list to West Point. The day after he graduated high school Borman received a letter telling him to report to the academy. He was in.

Three years later, Frank was on his way to Europe. His standing at the military academy was high enough for him to be chosen as one of a dozen cadets to tour Europe.

Borman arrived in Berlin just as the yearlong Berlin Airlift was coming to an end. "We flew into West Berlin on sacks of coal," he wrote later.[4] During the previous eleven months, the Soviet military had barred all ground transportation from entering an already struggling West Berlin. Food shipments were stopped. Coal supplies were blocked. Electricity, which came from a power plant in East Berlin, was cut off. With stockpiles for, at most, one to two months, it appeared that the 2.5 million inhabitants of West Berlin faced starvation unless the Western powers abandoned them to the communists.

The Soviets began the blockade not merely to exert their power. They had a legitimate fear of a re-united Germany, and felt that dividing the country would prevent the Germans from mounting another war against Russia.[5] They also wished to install a communist state in their East German zone. The presence of the capitalist island of West Berlin in the center of the communist zone made these goals difficult, if not impossible, to achieve.[6] When, on June 18th, the three Western powers unilaterally introduced a new West German currency, the Soviets responded in kind, further declaring that their East German marks were the sole currency for all of Berlin. "Russian legislation must apply to all sectors of Berlin,"[7] they proclaimed. The West answered this by bringing its new currency to West Berlin and announcing that both currencies would be legal tender there. General Lucius Clay, the United States Military Governor of Germany and commander of the U.S. forces in Europe, told the Russian Military Governor: "I reject *in toto* the Soviet claims to the city of Berlin." The Soviet reacted by cutting off West Berlin.

Clay's response to the blockade was a daring airlift, dubbed "Operation Vittles."[8] For the next eleven months planes landed in West Berlin every two and a half minutes, unloading powdered milk, flour, and diesel fuel, as well as the sacks of coal that Frank Borman had been sitting on. In the process, the people of Berlin accepted severe deprivation and near-starvation in order to resist Soviet rule. A second airport was quickly built, and at its peak the airlift was shipping more than 10,000 tons of supplies each day, including endless tons of coal needed to keep the people of West Berlin from freezing in that brutal winter cold.[9]

After almost a year, the Soviets finally realized that force would not get their former allies to leave Berlin. Furthermore, the siege had been hurting their own zone, which needed the shipments of coal, steel and machine parts that West Germany supplied. In May, 1949 the Soviets finally lifted the blockade, re-opening the rail lines and highways leading to Berlin.

The airlift continued, however, for another two months. When Borman arrived in June the Allies were aggressively re-stocking West Berlin with supplies, just in case the Soviets once again changed their minds.

Borman's European tour ended in Greece. There, a guerrilla army of communist rebels was trying (for the third time) to seize power by force. Knowing that they would certainly lose in the 1946 elections (some estimated they would only receive nine percent of the vote[10]), the communists abstained and declared war instead. Using bases in neighboring Yugoslavia, Albania, and Bulgaria, they made repeated forays into Greece, attacking villages and killing hostages.[11]

When Borman arrived in the summer of 1949, the rebellion was on the verge of defeat. The rebels had lost the support of Yugoslavia, and were defending their last strongholds within Greece itself. Borman and the cadets were taken to the front lines, where both sides were preparing for what in mere weeks would be the war's final battle.[12] En route, one of their convoy trucks hit a land mine, and once at the front the cadets watched for several days as the two sides lobbed mortar shells back and forth at each other.

Still young and eager to prove his mettle in the world, Borman had stood witness to the start of what was to be a forty year "cold war," a toe-to-toe stand-off which would dominate every aspect of the world's politics and culture. With the development in the late 1940's of the atomic bomb, the stakes rose to a frightening level, preventing outright war but forcing both sides to take actions that sometimes abrogated their own ideals. Machiavellian politics led to military dictators, the funding of terrorists, and indecisive military skirmishes throughout the world. In the end, however, the outcome of this stand-off determined whether the world's entire population would live under a state-run communist system or the free and chaotic capitalist system.

That 1949 journey through the ruins of Europe radically changed Frank Borman's perspective on life. His three years at West Point, dedicated to the motto of "Duty, Honor, Country," forged in him a desire not merely to fly airplanes, but to do it in defense of his country. The devastation of Europe and the communist oppression he saw there further committed him to the deeper principles he felt his country stood for: freedom, democracy, and the right of any human soul to pursue his or her dreams.

* * *

As Susan Borman notes today, "Frank Borman is the most uncomplicated man I have ever known." His passionate desire to dedicate his entire being to the military actually made him doubt the concept of marriage. To his straightforward mind, it had to be all or nothing. Only six months after arriving at West Point he wrote Susan a letter, explaining that he simply didn't have time for her anymore. With naïve innocence he had decided that he was going to live an obsessed, almost monklike existence in devotion to the cause of freedom.

Susan Borman, 1946.

Susan Bugbee was heartbroken. Up until that moment Frank had been the only man in her life. After crying her eyes out she decided to try and put Frank Borman from her mind. She began dating other high school boys.

Not more than three months later, Frank Borman realized how incredibly stupid he had been. He wrote Susan again, trying to repair their relationship. This time, however, she wasn't going to be so easy to get. While she didn't reject his offer outright, neither did she accept it. He had hurt her, and Susan had no intention of letting him hurt her again. Besides, she was now being wooed by a number of other boys.

For his sophomore and junior years at West Point Frank Borman did live like a monk, though not for his original reasons. He dated no one, and instead courted Susan by mail, sending her presents and gifts whenever he could. When he came back to Tucson during school breaks she was the first person he called. And though they dated, the relationship did not have its previous spark. Susan kept her distance. She wasn't going to be fooled again.

By the time of Frank's senior year in 1950 Susan was attending the University of Pennsylvania, studying dental hygiene. Several times he invited her to visit him at West Point, and she had gone. She still liked Frank despite everything, and could not make a clean break. Yet, she was also involved with another Pennsylvania student, and that relationship was starting to get serious.

Frank decided he simply didn't have a chance with Susan. After trying for two years to change her mind he had failed. It was time to move on and start dating other women. He called up a woman he had known in high school and asked her to come to West Point for a date. Not long after he gave her a ring and arranged for their wedding at the West Point Chapel when he graduated in June.

It wasn't right. In all the years at West Point he had not been able to get Susan out of his system. Within weeks he canceled the wedding, telling the woman that their fleeting engagement had been a big mistake. "I was a jerk," he admits humbly. He wrote Susan to invite her to come to his graduation.

She meanwhile had broken up with the dentistry student. She too couldn't get Frank Borman from her mind. Yet, when Frank asked her to come to his graduation but didn't mention anything about marriage Susan had finally had enough. She decided to take a big gamble. She told him "No." She went home to Tucson, hoping that by playing hard to get this last time she might at last get *him*. "I was terrified it wouldn't work."

Graduation at West Point was an important day in Frank Borman's life. And yet, he was unsatisfied. All the other graduates seemed to have fiancées. He was alone.

He knew now how much he needed Susan. As he and his parents made the long drive back to Tucson, he decided that he wouldn't take no for an answer, that merely getting together with Susan was insufficient. He was determined that she should be his wife.

Frank Borman was genuinely surprised how easy it was to convince Susan to change her mind and marry him. She only smiled slyly and said yes. She knew that the convincing had really been done by her.

July 20, 1950. Frank and Susan Borman on their wedding day,
Tucson, Arizona.

On July 20th, 1950, in a church in Tucson, Frank Borman and Susan
Bugbee became husband and wife, forming a partnership that was to last for
the rest of their lives.

## LOVELL

The waters of the western Pacific were cold and dark, and the night sky was
black. At 1,500 feet, Jim Lovell had no idea where he was, and had no way
of finding out. The instrument lights on his cockpit dashboard had failed
and his radio homing beacon wasn't working. Somewhere in that blackness
was his landing field, a tiny aircraft carrier only a few hundred feet long. If

Lovell failed to find this target, he'd have to ditch his plane and parachute into the bone-chilling waters of the Pacific.

The year was 1955, and Jim Lovell was making his first nighttime landing on an aircraft carrier over foreign waters.

As a child Lovell had been captivated by space and rockets. He would read the comic books of his time, showing Superman and Captain Marvel doing fantastic deeds, and he would draw his own imagined rockets and planes. He was mesmerized by Buck Rogers, Flash Gordon, and Jules Verne. And he would listen enthralled as his uncle, a navy pilot who had fought in World War I, told him stories of dogfights over the fields of France.

Fascinated with astronomy and space, young Lovell studied the stars and constellations. He read how astronomers had only recently discovered that the universe was much vaster than they had thought, comprised of endless numbers of grand galaxies.

Like Frank Borman, Jim Lovell's family was poor and struggling. His father, James Lovell, Sr., had been a coal furnace salesman in Philadelphia. When his father was killed in a car accident, Blanch Lovell suddenly became a poor widow with a twelve-year-old son and no means of support. She moved to Milwaukee to work as a secretary for her brother, who sold and marketed the same furnaces there. She and Jim settled into a tiny one-room apartment. The kitchen was in a closet, the beds folded up against the walls, and the bathroom was down the hall and shared by all the tenants.

Though they didn't have much, Blanch Lovell made sure that Jim had everything necessary to become whatever he wanted to be. By the time Lovell was seventeen, he had graduated from comics and books and was building his own model airplanes, flying them in an empty lot across the street from his apartment house. He and some high school friends even tried to build a homemade rocket. They had started out trying to construct a liquid-fueled engine, then switched to a dry-fueled solid rocket because it was easier and cheaper. They purchased gunpowder, packed it inside a cardboard tube so that it would burn instead of explode. For a fuse they used a soda straw filled with gunpowder and inserted into the rocket's tail end.

On launch day his mother watched from their apartment window, feeling both fear and pride. She could see her son and his friends in the lot across the street. She saw Jim prop the rocket against a rock, crouch down to light the fuse, then run for cover behind some nearby rocks. Seconds later the missile ignited, hurling itself high into the air with a high-pitched whistle and a bright flash. Then it exploded with a bang.

Also watching from Jim's apartment was fourteen-year-old Marilyn Gerlach. Earlier that year Jim, the sophisticated high school junior, had noticed this bright eyed thoughtful girl in the school cafeteria. Several times he had asked her if she would go on a date with him, but she had always said no. Though Marilyn thought Jim was good-looking and was very impressed that one of the school track stars was asking her for a date, he was so much older.

Near the end of the school year, Jim Lovell tried again. He had no date for his junior prom, and he wanted Marilyn to go with him.

Once again she equivocated. "Well, I don't know how to dance."

"Don't worry," he said. "I'll teach you."

For the next several weeks Jim brought records over to Marilyn's home and the two practiced dancing in her living room. Before long they were going steady.

At the same time that Jim Lovell was getting to know Marilyn Gerlach, he was also discovering that his fascination with rocketry might actually lead to his life's work. World War II had just ended, and a local museum had exhibited the V1 and V2 rockets of the just-defeated Germans. Staring at those formidable weapons built by engineers and scientists, Lovell suddenly realized that he would gladly spend his life building rockets.

He wrote a letter to the American Rocket Society, asking how he could become a rocket engineer. The society's president responded, explaining that "the whole field of rockets and jet propulsion is still so new that we do not know clearly what preparation is best for it." He suggested that Lovell get as thorough an education as he could, especially in fields such as thermodynamics or aerodynamics.

Now Lovell faced the same problem as Frank Borman. His mother didn't have the money to send him to college. He had applied to Annapolis but had been chosen as a third alternate, leaving him little chance of getting in.

Undeterred, Lovell took advantage of a Navy program called the Holloway Plan. The Navy would pay for him to get a two year engineering

Credit: Lovell

Jim Lovell and Marilyn Gerlach on board the U.S. Navy
sailboat *Freedom*, 1950.

degree, after which he would take fourteen months of flight training followed by six months at sea as an aviation midshipman. He would then begin a military career as a regular naval officer. Though this wasn't quite the same as a rocket engineering, the idea of flying advanced military airplanes appealed to Lovell almost as much.

For the next two years Lovell studied engineering at the University of Wisconsin at Madison, renting a room in a nearby family's house. And each weekend, Marilyn came by bus from Milwaukee to visit, staying in Jim's room while he moved to a YMCA near the campus.

At his mother's insistence, however, Jim applied a second time to the Naval Academy. She was afraid that when he went overseas as a midshipmen after his second year of college he would get caught in an overseas military action, and be unable to return to school for years afterward. He took her advice, and to his surprise this time he was accepted.

Blanch's foresight was almost clairvoyant. All of Jim's Holloway classmates ended up in Korea. Years would pass before they could complete their educations.

Lovell moved east, starting college all over again at Annapolis. By this time Marilyn was also going to college on a scholarship at Milwaukee State

Teachers College. When Jim headed to Maryland, he asked her to come east as well.

Marilyn had no doubt what she wanted to do. She gave up her scholarship, transferred to George Washington University in Washington, D.C., and got a part time job in an up-scale department store in Washington, selling woman's clothing.

She also typed Jim's school papers, including an astonishing twenty-four-page term paper on "The Development of the Liquid-Fuel Rocket." In this essay Jim described the early history of rocketry in the United States and Germany. He concluded enthusiastically that "The big day for rockets is still coming — the day when science will have advanced to the stage when flight into space is a reality and not a dream."[13] As she typed, Marilyn couldn't help feeling amused. "It seemed so farfetched," she later said.

They spent as much of their free time as possible together. Every Friday she traveled down to Annapolis, stayed with a local family, and joined Jim for the weekend socials.

And yet, while they had talked about marriage, Jim hadn't yet proposed, nor had they made any detailed plans about their future.

At the end of junior year the Academy held what was called the Ring Dance. At this formal ritual, the midshipmen received their class ring with the crests of both the Navel Academy and their graduating class embossed upon it. Should a midshipman be engaged at the time, his fiancee would be given a miniature of the ring at the same dance. During most of the dance the husband-to-be kept her ring in his pocket, while she wore his on a blue and gold ribbon around her neck.

At the center of the dance floor a short ramp led to a giant replica of the class ring, big enough for a couple to stand under. As the young couples danced, one-by-one they picked the moment to exchange rings. The couples dipped their engagement rings in a large bowl containing water from the seven seas, then walked together up the ramp and under the ring. There they placed the rings on each other's fingers, kissed, and thus became officially engaged.

By June of Jim's sophomore year he needed to pick out the style of ring he wanted. One weekend, Marilyn and Jim went to a jewelry store to look over the selection. As they both eyed the many styles, which included both the men's full size and the women's corresponding miniature, Jim Lovell nonchalantly indicated the miniature rack and said, "Well, which one would you like?"

Marilyn looked at him, a bit bewildered. "You mean you want me to have one of these?"

"Well, yes," he said. "Which style do you like?"

Rather than commenting on his unorthodox proposal of marriage, Marilyn merely looked at the selection, and made her choice.

\* \* \*

Then came the night in 1955, when Marilyn was back in California with their first two children, and Jim was sitting in the cockpit of his fighter jet, desperately trying to find his way back to the aircraft carrier.

After graduation and marriage, he had been assigned to a Pacific aircraft carrier group, training to fly jets at night. Though they never saw any MIGs, he and his squadron flew patrols and practiced night flying and night landings on their aircraft carrier. They also developed flying techniques for releasing their nuclear bombs and getting away as fast as possible.

And above all, they stayed prepared, ready to drop atomic weapons on their assigned targets in China and the Soviet Union should a nuclear attack be attempted against the United States.

The original plan this night had been for Lovell and three other planes to take off, fly at 30,000 feet for ninety minutes, lock onto a radio signal beamed from the carrier, and use this to rendezvous at 1,500 feet above the carrier deck. They'd then bring their planes home, one by one.

Lovell took off without problems, but after that nothing went right. First, the clouds rolled in, the fourth plane's take-off was aborted, and the ship told the pilots already in the air to forego flying at 30,000 feet and to return to formation at 1,500 feet. Locking onto the radio signal, Lovell began flying in the direction indicated. What he didn't know was that his instruments had locked onto a radio signal transmitted on the same frequency from an air base in Japan seventy miles away. As the other two pilots linked up over the carrier, Lovell found himself alone, gliding over the choppy waves of the Pacific ocean. And nowhere below him could he see the carrier.

Well, he thought, things could be worse. He swung his plane around and backtracked, scanning the ocean for any telltale sign of the aircraft carrier.

At this moment things got worse. In his youthful enthusiasm to improve his country's flying equipment, Lovell had improvised a small additional light for reading the tiny numbers printed on a reference card strapped to his lower leg. He now decided to turn this light on, and as planned, he plugged it into the instrument panel and flipped the switch.

Instantly he shorted out every light in his instrument panel. Now the inside of Lovell's cockpit was as dark as the outside.

Desperately he felt for a small penlight. Sticking it in his mouth, this was now all he had to illuminate his instruments. He could get his readings, one dial at a time, but it involved using one hand when he needed both to fly.

He turned the penlight off, having no idea what to do. As his eyes became accustomed to the darkness, however, he suddenly realized that he could see more details of the ocean surface. Below him was a faint greenish streak trailing out across the inky waves. As the carrier plowed its way through the ocean it churned up phosphorescent algae, illuminating the ship's wake in a dim gleam. By aiming for the head of the green wake, Lovell finally found his fellow wingmen. All three planes were once again together, circling above the carrier.

Now came the really hard part. Somehow, Lovell was going to have to land his plane on the tiny aircraft carrier deck while holding a penlight in his mouth.

He listened to the radio as the two other planes dropped to the carrier deck, snapping to a halt as the cables grabbed their tailhooks.

Now it was his turn. In order to hit the deck safely, he had to carefully lower his altitude from 250 to 150 feet just before he crossed over the deck.

In order to do that, however, he needed to read his instrument panel, and in order to do that, Lovell had to hold his penlight with his teeth while his hands flew the plane.

His first attempt almost rammed the side of the aircraft carrier. "Pull up, November Papa One, pull up!" the landing officer screamed at him.

"You're way too low!" Suddenly Lovell saw the side of the carrier loom up like a wall. With a desperate pull on the stick he ripped his plane away, barely missing the deck as he shrieked upward.

On his second attempt, Lovell decided that rather than risk a collision with the ship, he would drop down from 500 feet. With the ship screaming at him to lower his altitude, he cut his engines and plummeted like a rock towards that tiny carrier deck. Hitting the surface with a horrible thud, Lovell's plane bounced once, blew two tires, and then screeched to a jarring halt as the deck cables grabbed his tailhook.

When the first deck crewman opened his hatch, he looked at Lovell and calmly said, "Glad to see you decided to come back aboard."

Lovell found it hard to speak. "Yeah. Glad to be back."[14]


ANDERS


The horn went off suddenly, wailing a loud banshee cry throughout Keflavik Air Force Base. First Lieutenant Bill Anders and his radar operator had five minutes to get their plane skyborne, and all around him men climbed into battle array in anticipation of armed attack.

Radar had picked up an unidentified airplane approaching Iceland from the northeast, apparently coming from the Soviet Union and heading right towards what military people called the Air Defense Identification Zone. This zone, surrounding Iceland, was considered sovereign territory, and any intrusion by an unauthorized airplane could be considered an act of war.

Anders sprinted into the hangar, strapped his lifejacket on, and climbed into his cockpit. Gunning the engines of his F-89 Scorpion interceptor jet, now armed with over one hundred missiles, he pulled out onto the runway and screeched into the air.

The year was 1958, and it was Bill Anders's job to identify that unknown plane coming out of Soviet Russia and to destroy it if it had hostile intentions.

Born in Hong Kong to a Navy officer and a daughter of a Daughter of the American Revolution, Bill Anders had always been attracted to adventure and excitement. When he was four years old, his father had been assigned to the *U.S.S. Panay*, an American patrol boat policing the Yangtze River of China. As Arthur Anders patrolled the river, Muriel Anders and Bill would follow on shore, moving from city to city.

When the Japanese attacked Nanking, China in 1937, sinking the *Panay*, Muriel and Bill had to flee for their lives. The four-year-old was pulled from his bed to stand on a rickety wooden porch and watch bombs crash down several hundred yards from their hotel in Canton. He tried to run toward the explosions, but his mother grabbed him. The two soon boarded a steamer and slipped out to Hong Kong and on to the Philippines.

Lt. Arthur Anders, meanwhile, had also managed to escape. During the attack he had been forced to take command when the captain was badly injured. Anders, wounded in the throat and unable to talk, issued commands by writing them in pencil and even his own blood. No match for the Japanese, he refused to surrender, and ordered the crew to fight back. As the *Panay* sank, they fled with their wounded to shore and into the Chinese countryside, eventually escaping to Hankow. For his heroism Arthur Anders received the Navy Cross.[15]

The family returned to the States, where most of Bill Anders's youth was spent in San Diego. As he grew, the boy was attracted to science and engineering, but thought little about what he would do with his life. A soft-spoken man with a serious and proper demeanor, he hadn't developed a passionate interest in flying or rockets, like Lovell or Borman. All he knew was that he didn't want to do anything boring, and that whatever he did should conform to what he thought was right. Because his father had been a decorated Navy officer, and because his parents very much wanted him to join the Navy, he entered the Naval Academy at Annapolis in 1951. As a teenager Anders just assumed he'd become a ship's captain like his father.

It wasn't long, however, before Anders had second thoughts about spending the rest of his life on a ship. Each summer the academy sent the midshipmen off on what they called "cruises," joining a ship's crew at sea for several weeks in order to learn the ropes. In the summer of his first year Anders was assigned to a destroyer.

Always wanting to make the right impression, Anders put on his cleanest Navy whites and showed up about an hour and a half early. The officer of the deck checked him in, and told him to go to the stern and wait to be called.

Anders strolled to the stern of the ship, gazing at the ship's huge guns and armor. Suddenly he heard a loud hissing behind him, and he turned to see the ship's smokestack belch out a big black cloud of thick soot. He watched fascinated. Then he looked down and saw that his clothes were speckled with many many tiny bits of soot. Instinctively he brushed at them, and smeared the whole front of his whites. When a few minutes later he found himself being harshly chewed out for wearing a dirty uniform, the midshipman began to wonder at the justice of the Navy.

On the same cruise Anders attached himself to the ship's engine room chief, determined to learn all he could about maintaining a ship's machinery. A straight arrow who was generally uninterested in the typical shore activities of his fellow midshipmen, Anders would often forgo his leave to work with the chief. After several weeks of hard work, Anders had done so well that the chief put him in charge of the ship's throttle.

Then, with only ten minutes left on Anders' last engine room watch, the chief suddenly spoke up. "Mr. Anders, you're doing a good job. How'd you like a cup of coffee?"

Everyone in the engine room froze. A chief was in many ways the king of the engine room, the undisputed ruler of his realm. For him to offer midshipman Bill Anders a cup of coffee was to tell the ship that Bill Anders had now been knighted.

Anders nodded, and the chief filled a cup from his homemade jerry-rigged coffee machine and proffered it to Anders. "Here, you've earned it."

Anders, the proud but innocent young eager-beaver, smiled and naïvely asked, "That's great, chief. Have you got any cream and sugar?"

The chief looked at him in disgust. "Shee-it," he said, pouring the coffee into the bilge. "You'll never make it."

Anders wondered indeed if he could make it. He disliked playing politics, and it seemed that if he didn't play exactly the right politics at all times his career in the Navy would be difficult at best.

That same summer he went home to San Diego for a short leave, and went on a double date. Bill had had a crush on the younger sister of a high school buddy. Her name was Ann, and she worked as a pantry girl in a restaurant, making salads. Though he asked her for a date several times, she had always refused.

Finally she relented, but only if her brother Frank and another waitress named Valerie came along as well.

The foursome went to the beach. Ann and Frank soon found themselves abandoned while Bill and Valerie frolicked together in the surf.

Nineteen-year-old Bill Anders at that moment decided that he wanted to know more about sixteen-year-old Valerie Hoard. The next day he dressed up in his formal Navy whites and called upon Valerie's father to ask him if he could date his daughter. Henry "Casey" Hoard, a California highway patrolman, laughed, startled, and told him, "Sure, go ahead, she's dated every other boy in town."

Bill only had a few nights before his leave ended. The first night he took Valerie to the Starlight Opera. The next night they went to the Globe Theater to see Shakespeare. Each night he took her home and said goodbye with a handshake.

Valerie was baffled. She asked her mother, "What's wrong with him? He doesn't even try to kiss me." A typical California girl, she was a little taken aback by this serious, formal boy who seemed to approach life so seriously. She was more startled when she began to receive daily letters from him while he was on his summer assignment at sea.

He returned home at Christmas, 1952. Though his mother had arranged for him to attend all the local socials where he could meet the daughters of admirals, he ignored them all to date Valerie night and day.

By now Valerie was fascinated by this interesting guy who wrote long letters about philosophy and what he wanted to do with his life. Just before he left for Annapolis, he gave her a small pin, telling her to wear it as a souvenir.

When she wore it, however, Valerie suddenly discovered that she wasn't getting her usual number of dates. She hadn't realized that in a Navy town like San Diego, everyone recognized the pin as Bill's class crest. Her wearing it was equivalent to her advertising she was unavailable. *This is no good,* she thought. *I'm not having any fun!*

So while Bill thought they were going steady, she took the pin off so she could date other boys. Yet the two kept corresponding. By this time, Anders had thought about his future and had made some hard choices. First, he decided that he wanted to switch his commission to the Air Force. Flying combat jets seemed much more exciting than sitting on a ship.

Second, he decided that he wanted to marry Valerie Hoard. The Ring Dance was coming, and he began a full campaign to convince Valerie to marry him. When he came home that Christmas he drove her to the top of Mt. Helix, overlooking San Diego, and as the sun was setting formally proposed.

Valerie hesitated. She loved him and wanted to marry him, but this all seemed too fast. She was still only seventeen, was a straight "A" student in school, and wanted to go to college.

"You can't possibly say no to me," nineteen-year-old Bill argued. "If I get in the Air Force I'll get a two-month leave. It'll be the perfect opportunity for a honeymoon."

Upon returning to Annapolis, Bill's gentle but persistent campaign continued. New letters arrived daily, many including jewelry brochures showing Valerie a selection of engagement rings, all miniature variations of his class ring. She wrote back, describing how one ring, with a single small diamond surrounded by a cluster of even smaller diamonds, was the one she liked best. But she did not tell him to buy it.

At the same time she began taking instruction in Catholicism. Valerie had been raised as a casual Protestant. Her parents had left the choice of her church entirely up to her, and on Sunday her church of preference was usually determined by the church her friends attended.

Bill, however, had been raised a Catholic, and he took its sacraments seriously. If he was going to marry Valerie she had to convert. Though she hadn't yet committed to marriage, Valerie decided she had better find out about his religion, just in case. Twice a week she attended classes at two different churches, learning the theology of Catholicism.

Though she still wasn't sure she was ready for marriage, by June Valerie had decided to convert. She also knew that Bill had bought her a ring (though he wouldn't show it to her). And she had agreed to attend June week. She had never been to the East Coast, and the glamour of the Ring Dance seemed too

Credit: Anders

June 26, 1955. Bill and Valerie Anders on their wedding day
at the Naval Chapel, San Diego.

good to miss. She saved her money and flew across the country with all her
formal clothing.

Today she remembers this time with wistful good humor. "I was *so*
young." Each night they went to another formal dance, dressed to the nines.
Valerie wore crinoline skirts with long leather gloves—"much like Scarlet
O'Hara." On the night of the Ring Dance Valerie wore Bill's class ring on a
blue ribbon around her neck. At the center of the dance floor was the huge
replica of the ring.

Thoughout the evening Bill and Valerie danced. Finally the last shreds
of hesitancy faded, and Valerie knew that the time was right. They drifted
through the ring, and she took his ring from the ribbon around her neck and
placed it on his finger.

He took from his pocket a small box, and from it removed a miniature
ring with a single small diamond surrounded by a cluster of smaller diamonds.

He placed it on her finger, they looked at each other for one long silent moment, and then kissed.

Valerie had finally said yes.

\* \* \*

Three years later. Valerie was in San Diego with one-year-old Alan and one-month-old Glen, while Bill was an interceptor pilot stationed in Iceland.

He pulled up to his cruising altitude of 30,000 feet and headed southwest. For almost a year Anders had been patrolling the skies above Iceland, intercepting any unidentified aircraft that drifted into radar range. Most of the time they turned out to be commercial passenger jets flying slightly off course. During the winter, when the Iceland night lasted almost twenty-four hours, he would take off into a sparkling sky, the Northern Lights shimmering ghostlike above the horizon. The ground radar station would guide him to the plane, and his radar operator would shine a searchlight on its tail in order to read its number.

Now it was summertime, and the sun was bright and the sky clear blue. According to radar, the unidentified blip was coming from the northeast. Only Soviet military planes came from this direction, skirting Norway by flying over the Arctic Ocean. Anders turned due east in order to intercept the intruder before it reached Iceland.

Until he got close enough to actually see it, Anders had no idea what he was facing. If this blip heralded the beginning of a full-scale Soviet military attack, he would be fighting for his life. If it was a commercial jet he dared not misidentify it and shoot it down by mistake.

And if it was a merely a Soviet bomber on a routine mission, he would have to firmly escort it on its way, while simultaneously avoiding an international incident.

Ground radar guided Anders towards the intruder. When he was within twenty miles his radar operator picked it up on the F-89's own radar, and vectored the plane in from the side. Soon a speck appeared. The two planes were now less than 1,000 yards apart, and Anders could see that it was a Soviet bomber. Gingerly he swung around so that he was approaching it from the side. He had to make it clear to the other plane who he was and what his weaponry represented. He also had to make it clear that he would only attack if threatened. He knew that they had their guns trained on him as well.

Very carefully he eased up alongside, mere yards apart. Slowly he worked his way forward so that he was parallel with the Soviet plane's cockpit. He stared at the pilot and co-pilot, who stared back in turn. From what Anders could see, the Soviet bomber was alone and did not appear to be hostile.

For a few minutes the two planes flew in formation together as they headed west and out into the Atlantic. Then Anders noticed that his fuel was getting low, and asked his base for new instuctions. Knowing that the Soviet bomber was clear of NATO airspace, ground control directed him to break off, and he headed back to Iceland.

Before he did, Anders grinned at the Soviet crewmen and gave them the finger in farewell. The Soviets smiled back and held up a sign in English which read "We screwed your sister!"

Another day in the Cold War had ended.


## KHRUSHCHEV

On July 24, 1959, Nikita Khrushchev and Vice President Richard Nixon stood before an exhibit of an American kitchen in Moscow and, with news reporters and dignitaries crowded around, argued the merits of communism and capitalism. Khrushchev, a blunt undiplomatic man who never minced his words, looked at the display of ordinary American home appliances and scoffed, "I was born in the Soviet Union, so I have a right to a house. In America if you don't have a dollar — you have the right to choose between sleeping in a house or on the pavement."

Nixon responded, "To us, diversity, the right to choose, the fact that we have 1,000 builders building 1,000 different houses, is the most important thing . . . We do not wish to have decisions made at the top by government officials."[16]

Khrushchev looked Nixon in the eye and laughed. "We have existed not quite forty-two years," he said, waving an arm at the American exhibit.

"In another seven years we will be on the same level as America. When we catch up with you, in passing you by, we will wave to you."[17]

In the decade since Frank Borman had visited Berlin, the Cold War had only intensified, the Soviet Union emerging from World War II as an aggressive challenger to the Western World. And epitomizing that challenge was Nikita Sergeyevich Khrushchev, the First Secretary of the Communist Party and Premier of the Soviet Union.

Born in 1894, Khrushchev was a small, blustery man with a round face and round body. His father had been a poor man, struggling to survive first as a railroad worker, then as a farmer, and finally as a coal miner. When Nikita was thirteen he joined his father in the mines, apprenticing as a blacksmith.[18]

When the communists seized power in 1917 he quickly joined the revolution. Though his knowledge of communist and Marxist ideas was somewhat simplistic, "slogan-Marxism" as it were, he accepted its goals fully and with passion.[19] By the early 1920's he was fighting in the civil war against the White Russians. Soon after he was placed in charge of one of the very mines he had worked in years before. He subsequently rose swiftly through the party's ranks. In ten years he became a major political player under Stalin, surviving the purges of the late 1930's and instigating a few of his own when he was placed in charge of the Ukraine in 1938.[20] By 1953 he became head of the Soviet Communist party, and took full power in 1958.

Like many of his fellow communists, Khrushchev was never afraid to speak his mind about his faith in communism and his desire to see it dominate the world.[21] In November 1956 while attending a diplomatic reception in the Polish Embassy in Moscow, Khrushchev so eagerly proclaimed this faith that the envoys from a dozen Western countries walked out. "Whether you like it or not," he told them, "history is on our side. We will bury you!"[22]

Khrushchev believed that all power began with the state, and by doling out that power as the state deemed best a better society could be created.[23] "Centralization was the best and most efficient system," he wrote in his memoirs. "[Everything] had to be worked out at the top and supervised from above."[24] Such supervision included not only all economic activity and property rights but also all philosophical, religious, and intellectual thought.

The Western concept of liberty of conscience, where all citizens are free to choose their own religion or express their own personal beliefs, was alien to Khrushchev. Even when he allowed a "thaw," an easing of political oppression after the death of Stalin, Khrushchev placed limits on freedom.

> We were afraid the thaw might unleash a flood, which we wouldn't be able to control and which could drown us, [washing] away all the barriers and retaining walls of our society. . . . We wanted to guide the progress of the thaw so that it would stimulate only those creative forces which would contribute to the strengthening of socialism.[25]

Though dissidents and political opponents were no longer executed, many careers were still destroyed. Some were imprisoned, or exiled from the country. Books were banned, newspapers censored. And nothing could be published without government approval.[26]

Where religion was concerned, Khrushchev did not even allow a thaw. He saw religious expression as an evil that had to be stamped out. Soon after Khrushchev took power, a 1959 Pravda editorial stated that "religion is inimicable to the interests of the working masses . . . [hindering] the active struggle of the people for the transformation of society."[27] For the human race to rise to the new level, party workers had to strive for the "complete eradication of religious prejudices."[28]

These words signaled the beginning of an anti-religious campaign that in five years eliminated more than 10,000 churches in the Soviet Union, almost half of the entire country's places of worship. The property of believers was confiscated, priests arrested, and churches bulldozed, all in the name of atheism and rational thought.[29]

Khrushchev's faith in communism and the government's ability to guide society was also reflected in his unwillingness to tolerate political opposition. In 1953, soon after the death of Stalin, the East German government attempted to collectivize the country's farmers. The result was a revolt, nationwide strikes, and the sacking of the Communist Party offices in almost every major city. To regain control, the Soviet army rolled into East Germany, killing more than eight hundred people and re-establishing a sympathetic Communist government.[30] As

Khrushchev remembered, "Throngs of people went out into the streets. We were forced to move tanks into Berlin."[31]

In 1956, more than 200,000 people marched through the capitol of Hungary, demanding elections and an end to the Soviet occupation of their country. Khrushchev responded by sending the Soviet army into Budapest to smash the rebellion. "Once a movement like this gets started, the leadership loses the ability to influence the masses," he explained in his memoirs. The leaders of the Hungarian government were arrested and executed. A new government, loyal to the goals of the worldwide communist revolution as well as to the Soviet Union, was installed.[32]

Khrushchev's faith in communism did not manifest itself only in oppressive actions. He truly wanted to improve his nation, and made enormous sincere efforts to do so. As tough and as hard-willed as he was, Khrushchev also dreamed big and almost childlike dreams.

> There will come a time when our descendants, studying the heroic history of our deeds, will say: 'They did a great thing.' The people will wonder at how the workers of semi-literate Russia heading the working class went out to storm capitalism.[33]

Under his tempestuous leadership, the Soviet Union emerged from its borders to pose a powerful and real challenge to the West. In January 1958 Khrushchev's government agreed to the first extensive cultural exchange program with the United States, the climax of which would be two competing exhibits, one in New York City and the other in Moscow, both mounted in the summer of 1959.

The Soviet exhibit in New York City occupied the entire three floors of the Coliseum on 59th Street and Eighth Avenue. It included a full-scale three room apartment, a fashion show, Soviet-built cars, an electron telescope, a combination television and chess board, and numerous displays of technology, art, and culture. The exhibit extolled the glorious achievements of the communist way of life, as well as the hopes the Soviet leadership held for its people, its nation, and its philosophy.[34]

One Soviet triumph in particular seemed to prove Khrushchev's boasts. In the main entrance hall hung models of the four successful Soviet satellites, Sputniks 1, 2, and 3, and Luna 1. Sputnik 1 practically unhinged the American social order when it became the first artificial satellite on October 4, 1957. For three weeks it emitted regular beeps at standard shortwave frequencies as it circled the globe. Not only had America been beaten into space, Sputnik weighed six times more than the first American satellite, launched four months later.

Where space exploration was concerned, Nikita Khrushchev did more than just make fiery speeches. Khrushchev aggressively pushed Soviet engineers to design larger rockets, not just to build nuclear intercontinental missiles, but to embark on an ambitious space program that would bring communism to the stars. "People of the whole world are pointing to this satellite," Khrushchev said of that first Sputnik launch. "They are saying the United States has been beaten."[35]

Less than a month after Sputnik 1, the Soviets triumphed again with Sputnik 2, now five times heavier. More stunning to the American public, however, was that the capsule contained a dog named Laika, the first living creature to enter outer space. For about a week the world could hear Laika's heartbeat as she ate and slept. Then her oxygen ran out and she quietly died.

In May 1958 the Soviets launched Sputnik 3, a repeat of the first Sputnik mission, and then topped that feat by sending Luna 1 into solar orbit on January 2, 1959. This probe had been intended to hit the moon, and though the Soviet engineers missed their target by about 3,700 miles, their spacecraft became the first human object to escape the earth's gravity.

Meanwhile, the United States was having trouble getting its rockets off the ground. Just two months after Sputnik, with an entire world watching, the first American attempt to orbit a satellite exploded at launch. "Oh What a Flopnik!" was how one newspaper headline described the debacle.[36]

Though the U.S. was finally able to orbit six satellites in the next two years, fifteen other rockets were failures. Some exploded on the launch pad. Others broke apart in flight. Many simply fizzled, crashing back to earth.

Not surprisingly, the Soviets were proud of their lead in space. As Frol Kozlov, the First Deputy Premier of the Soviet Union, said at the opening ceremonies of the Soviet cultural exhibit, "We do not conceal that [these launches] required us to tax our strength considerably, but neither do we conceal our pride in the results of our toil."[37]

With the launch of the Sputniks, American society was panic-stricken: were capitalism and democracy unable to compete with a government-run communist state? More Americans then ever wondered if communism might actually be a better economic system.

\* \* \*

Frank Borman meanwhile had decided he needed a new challenge. From 1957 to 1960 he had been a teacher of fluid mechanics and thermodynamics at West Point. Now he applied and was accepted to the test pilots' school at Edwards Air Force Base. Though he was aware of the emerging space race, he was more interested in advancing his military career while also helping to develop his country's military defense. He, Susan, and their two boys, now nine and seven, climbed into a brand new 1960 Chevy and drove across the country. The Bormans were once again moving up in the world.

At the same time, Jim Lovell was accepted as test pilot at the U.S. Navy Aircraft Test Center in Patuxent River, Maryland, graduating first in his class. He had watched with growing frustration as all around him the science of rocketry bloomed. Lovell still wanted to build rockets, and flying Navy jets in routine patrols was no longer getting him closer to that dream. Pax River, as the test pilots called it, was one of the places he might get a chance to do so. He and Marilyn packed up their two kids, four-year-old Barbara and two-year-old Jay, and set off across the country for the coast of Maryland.

For the same reasons and at the same moment, Bill Anders also applied to the Air Force test pilots' school at Edwards. He had watched the momentum build in the space race and realized that human beings would soon be flying into space. "That's what I'd like to do," he told Valerie.[38] Unfortunately, because he lacked an advanced degree, Edwards Air Force Base rejected his application. Undeterred, he immediately enrolled in the Air Force Institute of Technology in Dayton, Ohio, and began two years of study in nuclear engineering and aeronautics.

Whether they knew it or not, all three men were putting themselves on the front lines of the Cold War, a war that was about to enter its hottest and most violent years.[39]

CHAPTER THREE

# "THAT EARTH IS SURE LOOKING SMALL."

To EVERYONE ON EARTH, the giant Saturn rocket and its Apollo 8 command module had now been reduced to three trebly voices on the radio.

To the astronauts, the earth had become a giant ball in space, shrinking from them at a startling rate. Its surface was a blue and white swirl of clouds and ocean, with some brown patches peeking out underneath the white. Only forty minutes after leaving earth orbit they were more than 20,000 miles from home.

With the earth still close by, Anders focused on getting 16mm movie and 70mm still shots of its retreating disk. He set the movie camera on its bracket, turned it on, then aimed a still camera out the window at the S4B engine. As he stared at the earth the spacecraft slowly shifted position, tilting so that the earth's globe moved from the bottom of the window to the top. Anders could see one of the third stage panels dropping away from them as it drifted out into space. The whole image reminded him of several scenes from the movie *2001: A Space Odyssey,* playing in the theaters at that very moment.

Only here, it was real.

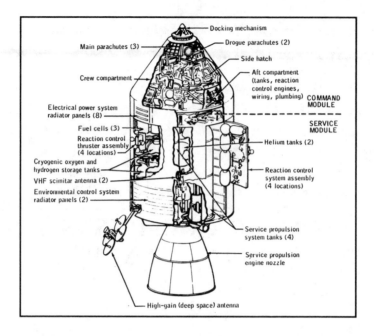

Other incongruities between that film and real life stood out. In *2001*, the spaceships floated through space to the melodic harmonies of Strauss and Stravinsky. Apollo 8's astronauts were serenaded by pop music, mostly records that Anders had given mission control prior to leaving Houston. At that moment the ground was piping up Herb Alpert and the Tijuana Brass.

The spaceships in *2001* were large, sophisticated, and designed to make space travel seem as ordinary as possible. Stewardesses brought passengers meals. Payphones were available to make private phone calls to family and friends. Everyone traveled in street clothes, and were provided magnetic soles for their shoes so that they could "walk" from one part of the ship to another.

Nothing on Apollo 8 was comparable. Apollo 8 was a small, experimental spacecraft, being tested with human occupants for only the second time. For its trip to the moon, it had two sections, the small mini-van-sized command module where the astronauts lived, and the service module, which held the main engine, power supply, and life support systems.

The command module was shaped like a very wide-mouthed ice cream cone, about twelve feet high with a round base about thirteen feet across. Its entire white surface was covered with a honeycomb made of fiberglass and injected with epoxy resin. When the spacecraft finally reentered the earth's atmosphere on its way home, this resin would burn off, taking the intense heat with it, and thereby protecting the three men inside.

Five windows had been built into that surface, two for Borman on the left, two for Anders on the right, and the round hatch window for Lovell in the center. Also built into that surface were two independent sets of six small thruster engines, each jet able to generate 94 pounds of thrust. One set of jets would be used for orientating the capsule as it reentered the atmosphere, the other reserved as a backup should the first fail.

Attached to the command module's base was the service module. In this thirteen foot long cylinder were oxygen tanks for supplying the astronauts with air, as well as three fuel cells, which combined the oxygen with hydrogen to generate electricity and drinking water.[1] On the service module's outside surface were four clusters of four additional rocket engines, used to adjust the spacecraft's orientation in space. Each one of these sixteen engines produced 100 pounds of thrust.

The spaceship's main engine, called the Service Propulsion System, or S.P.S. for short, was also part of the service module. The S.P.S., generating 20,500 pounds of thrust, was the rocket engine that would put the astronauts into lunar orbit in three days and, more importantly, blast them back to the earth when it was time to leave.

In this tiny spacecraft three men now drifted towards the moon. While Anders focused on photographing the earth, Borman piloted the spacecraft. Unlike driving a car, steering in space required more than left or right turns. Borman used two hand controls, resembling many of today's popular computer joysticks. One control accelerated the spacecraft in the desired direction, while the other merely pivoted the spacecraft around its center of mass. For example, by moving this second joystick backward or forward, Borman pitched the spacecraft's nose up or down. Tilted left or right, and the spacecraft rolled to the left or right. And twisting the joystick caused the whole spacecraft to yaw, a term borrowed from nautical dictionaries. Here the spacecraft was like a bottle lying on its side, and the pilot a teenager spinning it one way or the other, depending on the direction he turned the hand control.

As Borman maneuvered the spacecraft the abandoned third stage was causing him a lot of aggravation, trailing behind them in its own independent

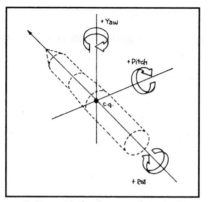

C.G. stands for "center of gravity."

path to the moon. "The damned S4B was uncomfortably close, its nose wandering within 500 feet."[2] Soon he was forced to turn the capsule away from the earth in order to keep watch on the booster.

Nor did he like how the S4B was venting fuel. "It's spewing out from all sides like a huge water sprinkler," he told the ground. "I believe we're going to have to vent or thrust away from this thing. We seem to be getting closer."

Moving away from the booster wasn't going to be as simple as Borman would have liked. The computers on the ground had calculated Apollo 8's heading, and determined that it was so accurate that the next mid-course correction was hardly needed. But mission control also wanted to fire the S.P.S. engine before the spacecraft got too far from earth. Like Borman's maneuvering controls, the S.P.S. was quite different from most earthbound engines, and was one of the reasons that NASA had gambled on sending Apollo 8 to the moon. The S.P.S. used hypergolic chemicals, meaning that when the fuel, a mixture of hydrazine and unsymmetrical dimethyl-hydrazine, made contact with the oxidizer, nitrogen tetroxide, the chemicals instantly ignited, producing thrust. No spark was needed. Without a complicated ignition system, the engine was simpler, and hopefully more reliable.

Two days before launch, however, engineers doing ground tests on another S.P.S. engine noticed an anomaly that posed a possible hazard. The engineers found that unless the combustion chamber of each new S.P.S. engine was primed, "wetted" with a small amount of fuel, there was a chance that the engine might explode the first time it was turned up to full thrust. The solution, not complicated, required giving the S.P.S. a single very short burst. The first mid-course correction, scheduled about eleven hours into the flight, would be the ideal opportunity to do this.

First, however, the ground engineers needed Borman to push the spacecraft a little bit more off course. The S.P.S. was too powerful an engine to make very small course corrections — it would be like using a bomb to kill a fly. By doing a sideways burn now with the capsule's small attitude thrusters, the spacecraft's course would be changed enough so that several hours hence they could use the S.P.S. engine to correct it.

In order to change the spacecraft's course, however, Borman needed to reorient the capsule, putting the earth in the windows instead of the S4B booster. "I don't want to do that," Borman explained. "I'll lose sight of the S4B." He and mission control compromised. The commander would position the spacecraft so that he could see both earth and booster, and make as much of a sideways burn as possible from this position.

Now however, Borman had to relocate the earth. For the next ten minutes he struggled, with Lovell's and Anders' help, to put both the earth and the booster in view. After five minutes Collins asked him if he had been able to do the burn. Borman responded, "As soon as we find the earth, we'll do it."

This brought a burst of startled laughter in Houston. It seemed absurd to say that the earth was hard to find.

Finally Borman was able to make the burn. He looked out his window at the S4B and reported, "We seem to be drifting away from this thing a little bit, although it is still pointing at us quite closer than I'd like."

Then he used the service module's side thrusters to put the module in what he called "barbecue mode," a slow roll spinning once per hour. This evenly distributed the burning heat of the sun over the entire surface of the spacecraft.

They had been in space for six hours, and awake for twelve. One by one the three men pulled off their bulky spacesuits so that they were dressed, not in street clothes, but in lightweight jumpsuits. Instead of magnetic-soled shoes, they wore cloth booties. And instead of "walking" from point to point, they simply pushed off one wall and floated across the cabin. Each man ate something, and things began to quiet down.

For a variety of reasons, astronaut Pete Conrad had pinned the nickname "Shaky" on Jim Lovell.[3] Though Jim *always* made things work in the end (like Borman and Anders, Lovell had never lost a plane in flight, and had finished ahead of Conrad in test pilot training), silly — and sometimes life-threatening — things seemed always to happen around him.

Shortly after reaching orbit, Lovell started to move from his couch to his navigation station in the lower equipment bay. As he did so he accidentally pulled on the toggle switch for his life vest, activating it. Suddenly he wearing two bulging and growing balloons in a space that gave him very little room to manuever.

Borman and Anders laughed. When Borman spoke to the ground, he described Lovell by saying that "we've got one full Mae West with us."

Since the vest was filled with carbon dioxide, deflating it would cause the excess $CO_2$ to saturate the filters for cleaning the capsule's atmosphere. And Lovell had to get rid of it if he was to do his work.

Lovell carefully glided to the urine dump. Normally an astronaut would insert his personal plumbing into a hose and void his liquid waste into the great emptiness of space. Now Lovell inserted the hose into his life vest, squeezing the carbon dioxide gas through it and out of the capsule.

Soon he no longer resembled a big-breasted Hollywood star, and could store his spacesuit with the others.

With Borman steering and Anders alternating between taking photos and monitoring the capsule's operations, Lovell now got busy doing his main task, trying to prove that a human being could pinpoint his position in space without the use of ground-based help.

With the spacecraft's navigational telescope and sextant, Lovell sighted on several stars as well as the horizon of the earth, using this data to triangulate the spacecraft's course and position. This manual navigation system had been designed as a backup to ground-based calculations. Should communications fail, the astronauts would then use Lovell's sightings to program their course changes by hand.

Called the inertial measuring unit (or I.M.U.), this equipment tracked the spacecraft's orientation relative to the earth and solar system. Because space has no up and down or horizon line, the astronauts needed some other reference for pointing the spacecraft's nose and engines in the right direction.

Set on three gimbals, one for each of the three dimensions, the I.M.U. was held stable by gyroscopes. Beginning from its initial setting on the launchpad, the unit recorded any changes thereafter of the capsule's orientaton. These changes in turn were reflected by a control panel indicator the astronauts called the eight ball, a grapefruit-sized sphere incised with the sky's longitudes and latitudes. As the capsule pitched, rolled, or yawed relative to the stars, the I.M.U. told the eight ball to turn and roll correspondingly.

Originally NASA had planned to shut the I.M.U. off when not in use in order to conserve power, and let Lovell re-set it manually prior to each burn. Borman and the other astronauts disagreed vigorously with this idea. As Borman wrote later, "Experience had taught me that when you have something running perfectly, particularly a mechanical or electrical device, it's best to leave it alone."[4] After much discussion the engineers agreed. The I.M.U. was left on for the entire flight, with Lovell's sightings used merely for back-up.

This decision seemed even wiser now, only a few hours from earth. While still in orbit Lovell had found it difficult to locate any stars because of the planet's glare. Now, on the way to the moon, the venting from the S4B booster was making star identification tricky. The fuel scattered into millions of tiny frozen globules, and the light reflected off these particles to fill the void around them with many "pseudo-stars."

Nonetheless, Lovell made an attempt at pinpointing his location, and radioed his results back to earth. There, ground controllers compared his results with their own to see how accurate the figures were.

Accuracy was essential. Not only were the astronauts traveling farther than anyone ever had at a greater speed, but the elements that made up their motion were exceedingly complex. The spacecraft had left a planet whose surface was moving at about 1,000 miles per hour as the globe rotated. That planet was also cruising through space at 67,000 miles per hour. The spacecraft was aimed at a moon moving at 2,300 miles an hour relative to the earth, with an orbital plane that differed from the spacecraft's. Each of these vectors had to be incorporated into both Lovell's and the ground engineers' calculations so that they could aim Apollo 8 not at where the moon was, but at a point in space it would reach three days hence. And their calculations had to be accurate within four ten-thousandth of a single percentage point.

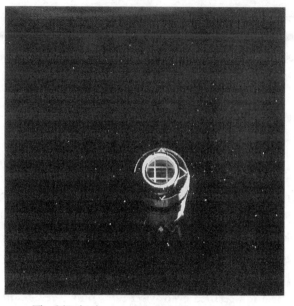

The S4B third stage, surrounded by pseudo-stars.

This was not unlike a person jumping from a speeding roller coaster car and trying to catch a bullet shot past them as they fell.

About ten hours into the flight, Mike Collins's shift ended and he was replaced at communications by Ken Mattingly. Mattingly had joined NASA in April 1966 as part of the fourth class of astronauts. Addicted to flying, he had spent his life in the Navy finding ways to get himself in the air. Now he was on the Apollo 8 support team, handling radio communications with a spacecraft almost 60,000 miles from home. To Mattingly this seemed an exhilarating turn of events.

He now radioed that he had the numbers for the first mid-course correction, scheduled for one hour hence. Borman, still at the controls, told him he was ready to take them down.

This aspect of early space exploration should horrify a modern computer user. Apollo 8's on-board computer was not capable of doing the calculations necessary for each planned rocket firing. Nor was its programmable memory of approximately four kilobytes (about eight to thirty thousand times smaller than today's average desktop) large enough to store much data.[5]

The ground computers did the calculations, and then mission control verbally passed the numbers up to the astronauts. They in turn then manually entered this data into the computer, which in turn controled the automatic firing of the spacecraft's engines.

Passing the numbers up from the ground, however, was hardly a simple task. Consider the list of numbers that Ken Mattingly now radioed to Frank Borman for mid-course correction number one: "Okay. Sixty-three thousand, one hundred and forty minus one sixty-three, plus one twenty-nine zero thirteen fifty-six forty-eight ninety-seven, minus zero zero five, ninety-nine, plus zero zero zero zero zero, plus four seven zero one six, one seventy-seven one forty-three zero zero zero November Alpha, plus zero zero one ninety-seven forty-seven zero twenty-five fifty-one four sixty-eight eighteen twelve twelve eighty-three two fifty-seven zero twenty-three."

He took a breath, then continued. "Up two sixty-three, left seventeen, plus eleven ninety-five, minus one sixty-five zero zero one twenty-six eighty-three three fifty-six zero eight zero fifty forty-seven zero five, north stars, zero sixty-eight zero ninety-seven three fifty-six, no ullage."

Borman, who was writing this litany down as he heard it, now repeated it back to Mattingly, confirming that he got it right. Later, Jim Lovell entered the numbers into the on-board computer, which would then be programmed to fire the rockets when scheduled.

An hour later and eleven hours into the flight, the computer did exactly that. As commanded, the S.P.S. engine fired, burping for just over two seconds. Not only did this blast successfully prime the S.P.S. engine, it was so accurate that it made the next two course corrections unnecessary. Mission control decided that Apollo 8 could continue on its course to the moon, still three days away.

* * *

Valerie Anders was practically a prisoner in her own home. The mob of reporters on her front lawn had grown so large that she didn't dare go outside. To her chagrin her two older boys, eleven-year-old Alan and ten-year-old Glen, couldn't resist talking with the reporters and getting their pictures taken.

As with every astronaut wife, she had been assigned a NASA press liaison to schedule press conferences. Soon after the spacecraft left earth orbit she went outside to answer questions, giving them what she called the standard "courageous astronaut wife" answers. Nonetheless, her own exuberance came out. "It was about the greatest thing I've ever seen," she told reporters.[6]

She spent the rest of the day in her home. By that Saturday evening Valerie had to get away. Her next door neighbors, astronaut Charlie Duke and his wife, Dorothy, were holding a Christmas eggnog party to celebrate the completion of their new house. Duke had been selected as an astronaut in 1966 and was now settling into the Houston community. Leaving her kids in the charge of her au pair, she snuck out her back door and slipped across the driveway into the Dukes. Once there she joined the party for an hour or so, drinking eggnog and chatting with Mike Collins and Jerry Carr about how well the flight was going.

Then she went home to put her kids to bed and go to sleep. As she lay in bed alone she listened to a special "squawk box" that NASA had installed in her bedroom. This box, placed in all the astronauts' homes, was linked directly to mission control, and allowed her to hear the ground-to-capsule communications.

Valerie had always been a good sleeper, and had thought the astronauts' voices would lull her to sleep. Instead, their talk fueled the high she had been on all day, and she lay there listening with endless interest.

Finally she turned the box off. In the silence of her bedroom she drifted quietly to sleep.

* * *

In Florida, Marilyn Lovell had watched the launch with exhilaration. She held her youngest son Jeffrey, almost three, and tilted her head back as the Saturn V climbed into the air. As Jim had predicted, the rocket slid slightly to the side before it cleared the tower, then rose majestically on a pillar of smoke and fire. Then the sound wave hit her, and the noise was so loud it sounded like the staccato distortion on a overloaded sound speaker.

Because she had watched other live launches with Jim, including the Apollo 7 launch two months earlier, she was prepared for the experience. All she could think about as she watched that rocket rise was how happy her husband must feel, doing what he had always dreamed of doing.

Afterward, the NASA press officer took her to a nearby motel for a poolside press conference, a task she found far more nerve-racking than watching the launch. She was astonished when a reporter asked if Jim was going to name a mountain after her. *How on earth did they find out about that?* she thought to herself. She told the reporters to ask Jim about it when he got into lunar orbit. She and the kids then returned to their beachside cottage, where a crowd of friends and relatives had gathered with food and champagne.

By the end of the day she was exhausted. As she lay in bed, she could hear the gentle crash of the ocean waves on the beach. But she had no squawk box, and was out of touch with the one voice she wanted to hear most. Not surprisingly, Marilyn found it difficult to sleep that night.

* * *

By Saturday night the three astronauts had been awake for almost nineteen hours. Frank Borman passed the controls to Bill Anders and went to bed. He slipped into the small space under the three couches and slide into a thin baglike hammock, designed to hold him in place as he slept.

After two hours of sleeplessness, however, he decided to let the ground know about his wish to take a sleeping pill. As was the custom, anytime the astronauts wanted to take any medicine or pills they first cleared it with the doctors on the ground.

The sleeping pill didn't help. Borman dozed, sleeping fitfully.

For five hours Lovell and Anders traded off as pilot. Neither said much. The spacecraft was on course, everything was working, and it was the middle of the night in Houston. It was time for some quiet.

By 1 AM Borman found that not only couldn't he sleep, but he was getting increasingly nauseous and queasy. Suddenly he was retching his guts out in the lower level of the command module. Floating balls of vomit drifted throughout the cabin. Anders watched in fascination as one particularly large blob, pulsating from spherical to ellipsoid as it glided past, gently splattered itself across Lovell's chest. To avoid the sudden stench, Anders fastened an emergency oxygen mask over his mouth and nose.

Meanwhile Borman was also having an attack of diarrhea, and Lovell and Anders found themselves scrambling about the cabin, trying to capture blobs of feces and vomit with paper towels. So much for the glamour of space flight.

At first Borman was reluctant to bother the ground with the problem. He already felt better, and was convinced that his illness had been caused by the sleeping tablet he had taken. He took the controls so that Lovell and Anders could take their rest breaks. Both had been up for almost twenty-four hours.

Now Anders found that he couldn't sleep. Besides the unpleasant fact that tiny bits of vomit and feces would periodically drift by, he found it difficult to rest in zero gravity. As the flight's rookie, he missed putting his head on a pillow, and every time he began to drift off he would jerk awake, spooked by the feeling that he was falling.

Though Lovell slept better, having been in space twice before, he only got five hours of sleep. As he told Mike Collins that morning, "Oh, you know. The first night in space . . . it's a little slow."

By 7 AM Sunday, the astronauts were all up. Lovell and Anders now finally convinced the commander that they should tell the ground about his stomach problems. To avoid announcing his diarrhea to the entire world, however, the astronauts recorded their report on the in-cabin tape recorder. These tapes were periodically dumped at high speed to the ground and could be reviewed by mission control in private.

Still, they didn't have any way of telling the ground that it was *very important* to listen to the tapes. All they could do was hint broadly that the tape contained some interesting material. Borman asked mission control if they had been reading the tape dumps. "How's the voice quality been?" he asked. Anders suggested that mission control carefully evaluate the voice comments. "You might want to listen to it in real time to evaluate the voice," he added twenty-five minutes later when he realized that no one on the ground had yet gotten the hint.

Finally, Houston caught on. Flight Director Charlesworth decided to go to a back room and listen to the tapes. Then he called in Charles Berry, NASA's medical director, to listen as well. They were immediately worried that Borman's illness had been caused by the Van Allen radiation belts that Apollo 8 had passed through on Saturday. Though sensors indicated that the total radiation experienced by the spacecraft appeared to be less than that of a single chest X-ray, no one was exactly sure what effect the belts would have on humans. If all three astronauts began to have the same symptoms, it might become necessary to abort the trip to the moon and bring the astronauts back quickly.[7]

Mission control actually had two different but identical control rooms. Because the third floor control room being used for Apollo 8 was packed with engineers, visitors, and reporters, all of whom could eavesdrop on the ground-to-capsule communications, Charlesworth, Berry, Mike Collins, and a handful of other individuals moved down to the unused and empty control room on the second floor. There Collins radioed the spacecraft on a private line, and he and Berry discussed the situation with Borman, who not only insisted that he felt fine, but that the other two men felt okay as well. Lovell and Anders had felt a little queasy when they had first gotten out of their spacesuits, but that had passed quickly.[8]

Borman assumed that he either had had a case of the twenty-four hour flu, or that the Seconal tablet had upset his stomach. All agreed that the mission to the moon should go on.

* * *

That same morning, Susan Borman and her two sons, seventeen-year-old Fred and fifteen-year-old Ed, went to Sunday services at St. Christopher's Episcopal Church.

She hadn't slept much herself, and since launch had attempted to numb her mind to the unfolding events. Soon after blastoff on Saturday she held a press conference on her front lawn with her sons, her husband's parents, and the family dog Teddy. "I've always been known as a person who had something to say," she told reporters. "Today I'm speechless." To ease her tension she introduced Teddy to the reporters. "He was right with us all the way," she joked.[9]

Soon her home was like an open house. People bustled in and out, bringing food, chatting, listening to the squawk box or watching the television together. When Susan needed time alone she went into her bedroom and closed the door for an hour or so, trying to sleep.

About 9:00 PM the house began to clear out and quiet down. Then Susan sat and nibbled on food, had a drink, smoked a cigarette, and listened to the babble between the ground and the astronauts. It relaxed her to hear them talk about routine matters, since this meant that the expected disaster had not yet struck.

She wanted to go outside and look at the moon, but it was cloudy in Houston. While commentators on the television were talking about how people across the nation were gazing at the moon in wonder, she felt frustrated at her inability to see it.

Very early Sunday morning she got a call from the NASA doctors, telling her about Frank's sickness and asking her for her opinion. Did this happen often? How did he handle such illness? Did she think it would incapacitate him?

Susan laughed and told them, "So what? What's the big deal?" Of all her worries, Frank was not one of them. She knew, as she had always known when he was a test pilot, that he would never be the cause of any problem. *Frank would always get it right*, she thought. Her fear was that the equipment would fail.

* * *

Marilyn Lovell meanwhile had returned to Houston, flying back early Sunday morning with the kids.

Almost immediately her home filled with neighbors, astronauts, and wives, all bringing food. She found it hilarious how many showed up at her door with trays of deviled eggs.

At 2:00 PM Sunday afternoon the astronauts were scheduled to hold their first televised press conference in space, preempting Sunday football. By now their tiny spacecraft had climbed more than 138,000 miles into the sky, and though the earth's gravity was steadily pulling at them, they were still moving at more than 3,000 miles per hour.

The black and white camera was small for its day, about the size of a large hardcover book and weighing four and a half pounds. Very similar to the one used on the first manned Apollo mission three months earlier, these were in a sense the world's first hand-held video cameras.

Borman, who had adamantly fought to keep the mission as simple as possible, had tried to keep the camera off as well. He argued that the extra weight was unneeded and that the extra chore of televising press conferences would only distract the astronauts.

Borman lost the argument. NASA very much wanted to give the people on earth a personal view of the first human flight to another world. The camera was included, and six separate space telecasts were scheduled, two on the way to the moon, two in lunar orbit, and two on the way home.

The first conference began with Bill Anders as cameraman, shooting Frank Borman floating freely in the cabin. Anders then worked his way down to what the astronauts called the lower equipment bay, a small area below their feet where Jim Lovell was supposed to be doing navigational sightings. Instead, he was preparing himself a snack. "We gotcha!" Borman joked.

"This is known as preparing lunch and doing P23 at the same time," Lovell grinned. P23 referred to program 23, a computer routine Lovell used when he did navigational work. Rather than do this, he instead demonstrated to his earth audience how he injected water into a bag to make chocolate pudding.

Next Anders took off the wide-angle lens and put on the telephoto lens so that he could show the earth-bound what their planet looked like from 138,000 miles away. Unfortunately, the telephoto lens wouldn't work, and the normal lens only showed what Ken Mattingly called "a real bright blob on the screen."

"I certainly wish we could show you the earth," Borman lamented. "It is a beautiful, beautiful view, with predominantly blue background and just huge covers of white clouds."

Bill Anders and toothbrush. The command module's instrument panel is on the left, the couches on the right. The floating cable is the power and transmission line for the video camera.

Failing at this, they turned the camera back inside, and with Borman acting as cameraman Bill Anders used his toothbrush to show how things floated in zero gravity. "He has been brushing regularly," Borman noted.

With one last close-up of Jim Lovell to "let everyone see he has already outdistanced us in the beard race," the astronauts signed off. Though their first television show had lasted barely fifteen minutes, and had failed to show the earth to the people on the ground, it had served as a powerful teaser.

Lovell's last words before they turned off the camera were "Happy birthday, Mother!" He hadn't forgotten that this was her 73rd birthday, and he knew she would be watching in Florida.

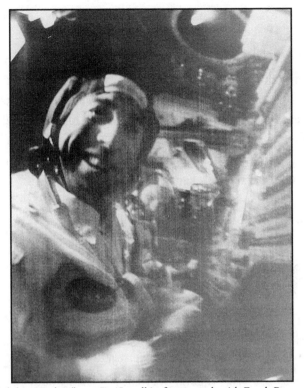

"Happy Birthday, Mother!" says Jim Lovell in foreground, with Frank Borman behind him.
The caps the astronauts are wearing were called "Snoopy hats" because
they resembled the cartoon character's helmet when he pretended
to be a World War I flying ace.

Blanch Lovell was delighted. "I just can't get over it," she later told
reporters. "When they had so many things to do in space that he would think
of his mother on her birthday."[10]

In Houston, Marilyn smiled. She knew how glad it would make Blanch
feel to hear Jim say this from space. *He's always thinking of his family,* she thought.

During the telecast Marilyn suddenly became conscious of something
else as well: for the first time it dawned on her how far away Jim was going.
*They really are heading to another world,* she thought.

\* \* \*

Valerie Anders, watching from home, was entranced. The three men seemed to be having so much fun bouncing around in zero g. Valerie also could see that whatever had made Frank Borman ill couldn't have been that serious. By now his nausea had become public knowledge, and much was being made in the press about the dangers of "space sickness." As she told reporters a short while later at another front lawn press conference, "They wouldn't be fixing chocolate pudding if they felt bad."[11]

Susan Borman also watched the broadcast, but to her the only thing of importance was to keep her guard up, to be ready to react properly when disaster struck.

\* \* \*

For the next nine hours the astronauts had very little to do other than some basic housekeeping chores. All three made several attempts to sleep, with mixed results. With Houston's approval, Borman had decided to shorten the rest periods and make them more frequent. This would not only make the time seem to go faster, but it might improve their chances of sleep.

Anders in particular found sleep difficult. He couldn't help worrying about the spacecraft and its operation, both of which were his direct responsibility. Even as he lay there, supposedly sleeping, he watched the others, checking them to make sure they were pressing the right buttons and flipping the right switches. At one point he thought Lovell was reaching for the wrong switch, and startled him with a correction. Lovell had thought Anders fast asleep.

Finally, however, Anders did doze off, if for no other reason than that the reality of space flight was beginning to settle in: simple boredom. They had a long way to go, and nothing but a lot of repetitive work to do until they got there.

It is difficult to imagine the distances involved. They started their journey at a speed of over 24,000 miles per hour, fast enough to have taken

Ferdinand Magellan around the globe in one hour instead of three years. Even so, the astronauts needed three full days to get to the moon. After forty hours, moving faster than any human in history, they still had thirty more hours of travel time left.

In the meantime, all they could do was maintain the spacecraft's systems and wait. And watch the earth steadily and relentlessly shrink behind them.

By 11 PM Sunday night, their isolation from earth was truly beginning to have an impact. Earlier that day, Mike Collins had given the astronauts what would become his daily morning news summary, a little report of the world's headlines, intended to keep the spacemen in touch with earth. Unfortunately, Frank Borman had missed that first news summary because he had been too sick to pay attention.

Now, at 11 PM, with Lovell and Anders trying to sleep, Jerry Carr was at capcom. He decided to fill Borman in on what he had missed. Carr had been a Marine fighter pilot and had both a bachelor's and master's degree in aeronautical engineering. When NASA had announced in 1964 it was looking for more astronauts, Carr had applied more out of a passing curiosity than any real expectation of acceptance. He was astonished when he was picked for the program in 1966. His initial astronaut training complete, he was assigned as support crew to Apollo 8.

Besides giving Borman the football scores, however, Carr also noted that the only real news (outside of the Apollo 8 mission) was that the crew of the *Pueblo* had finally been released. Eleven months earlier, the *U.S.S. Pueblo* and its eighty-two crew members had been seized by North Korean gunboats. The ship, a reconnaissance vessel doing surveillance in international waters, had been boarded and one man killed. The rest of the crew was taken to a prison camp in North Korea.[12]

The crew had been tortured and forced to issue false public confessions. Now, in order to obtain the release of the men, the U.S. negotiator had agreed to sign a document admitting to U.S. espionage in North Korean waters, even though the negotiator called the document a lie even as he stood there signing it. "Apparently the North Koreans believe there is propaganda value even in a worthless document," said Secretary of State Dean Rusk.[13]

Carr described to Borman how the *Pueblo* crew was released. "It took about thirty minutes for all eighty-two men to come across the Bridge of No Return, that's the one separating North and South Korea. . . They brought the body of the crewman that was killed also."

Then Carr gave Borman some football scores, and soon the conversation drifted to the weather. "It's beginning to feel like winter again," Carr noted.

"Good time for Christmas," Borman mused. "Good weather for Christmas."

Carr felt a need to talk some more. "Frank, we had a little eggnog over at Charlie Duke's tonight. Val Anders dropped by. She's looking fine. Tell Bill she's doing real fine."

"Fine." There was a long thirty second pause, and then Borman spoke up. "How do you like shift work, Jerry?"

"It's great, Frank. You've got the Black Watch watching you tonight." The Black Team was the official name for Carr's shift, because they were scheduled to work mostly late night hours.

"Yes, that's what I figured." There was another long pause, this time lasting more than two minutes. Borman sat and stared at his tiny home planet, far, far away. The conversation with Jerry Carr had made him think of Susan and his family. He struggled to think of other things.

He thought of the *Pueblo*. Borman felt, as did many military men, that the *Pueblo*'s captain had surrendered too easily the previous year. Anders, for example, remembered how in China his badly wounded father and his crew had refused to capitulate, fighting until the *Panay* was sunk.

Borman thought of the *Pueblo* crew's imprisonment, torture, and release, and how those men would now be able to celebrate Christmas with their families. He stared out the window. He could almost feel the vast black emptiness of space that surrounded the earth. The vastness seemed to press down on him like a terrible weight. He exhaled. "Boy, Jerry. That earth is sure looking small."

Carr could only agree. "Roger. I guess it'll get smaller too."
Sunday, December 22nd, 1968 was coming to an end.

## CHAPTER FOUR

# "WE STAND FOR FREEDOM."

## BERLIN

AT ABOUT 1 AM ON THE NIGHT OF AUGUST 13TH, 1961, the streets of the Soviet zone of Berlin were filled with the roar of vehicles. Hundreds of East German trucks, escorted by Soviet tanks, were on the move, converging on the perimeter of the American, British, and French zones of West Berlin.

At the same time, all subway trains attempting to cross the border between East and West Berlin were stopped, their passengers forced to disembark and find other ways home. Even the trains that only cut through East Berlin on their way from one part of West Berlin to another were emptied of passengers before being allowed to proceed. Announcements from loudspeakers blared that subway "traffic will be interrupted until further notice."[1]

At the Brandenburg Gate, just inside the Soviet zone, the six large floodlights that illuminated the plaza from the Soviet side abruptly went dark, and in the dim streetlight a single truck sped between the gate's pillars to deposit a dozen machine gun-armed East German soldiers. Behind them came additional soldiers, carrying barricades, soon followed by an almost unending line of military trucks. From these the soldiers unloaded heavy eight-foot high concrete posts and rolls of barbed wire and wire fencing.

Bulldozers and heavy construction equipment appeared. Silently ignoring the insults hurled at them by the small crowd that had gathered on the West Berlin side, the soldiers began drilling holes in the ground. Soon they eased the concrete posts into place and attached the wire fencing to them. Behind this wire wall they then unrolled the bushels of barbed wire, creating a second, more deadly obstruction. As they worked, the line of military trucks kept rumbling into the plaza, with more soldiers quickly unloading more concrete posts and wire.

By dawn the East German soldiers had built a fence 2,500 feet long across the face of the Brandenburg Gate. On the radio the East German government announced:

> In the face of the aggressive aspirations of the reactionary forces of [West Germany] and its NATO allies, the Warsaw Pact member states cannot but take necessary measures to guarantee their security and, primarily, the security of [East Germany] in the interests of the German peoples themselves.

> They were therefore establishing controls

> on the borders of West Berlin which will securely block the way to the subversive activity against the socialist camp countries.[2]

This "subversive activity" referred to what in recent weeks had become an unceasing flood of East German refugees, fleeing to the West by entering East Berlin and taking the subway across. When rumors indicated that this escape route might soon be closed, the numbers of refugees skyrocketed to over 30,000 in July and almost 20,000 in the first twelve days of August. This exodus of East German citizens had made the communist state one of the only nations in the world with a declining population. In the twelve years since Frank Borman had seen those East German refugees in the Dachau camp, 2.8 million people, seventeen percent of the total East German population, had fled Khrushchev's socialist paradise.[3]

Now that tide was to cease. Two weeks before, Walter Ulbrecht, East Germany's President and Communist Party chief, had come to Moscow demanding that Khrushchev and the Soviets help him stop the flow. Together the two rulers decided that the solution was "the establishment of border control," as Khrushchev euphemistically called the construction of the Berlin Wall.[4] Khrushchev, like Stalin, still wished to see East Germany succeed as a communist state, and like Stalin, he had concluded that the only way to make this happen was to restrict the freedom of Germans to travel.

Controlling the movement of citizens was an important priority for Khrushchev's government. In the Soviet Union if a person wished to relocate from the town of their birth, authorization was required, and indicated on a citizen's passport. In the campaign to snuff out religion, now running at full speed, K.G.B. officers confiscated the passports of priests, and demanded that they leave the town and church to which they ministered. If a clergyman refused, the K.G.B. would arrest him and prosecute him for violating the passport regulations. At monasteries across the Soviet Union religious clerics were being arrested and jailed. One priest was condemned three different times, serving three and a half years of hard labor from 1962 to 1966. Each time he was released from prison he returned to his monastery, and each time the K.G.B. re-arrested him.[5]

Now Khrushchev moved to apply this same standard to East Berlin. He obtained a map of Berlin and he and Ulbrecht sat down to work out the details. "It was a difficult task to divide the city of Berlin," he reminisced in his memoirs. "Everything is intertwined. The border goes along a street, so one side of the street is East Berlin while the other is in West Berlin."[6] After much discussion, the two communist leaders "decided to erect antitank barriers and barricades."[7]

By Sunday night, East German guards were patrolling that barricade with machine guns and tear gas. At Teltow Canal, which also formed the border but where no wire fence had been built, many refugees escaped by swimming across its short width. By Monday, the border guards moved in, and when a young couple dove into the water, the guards opened fire. Though the couple escaped unharmed, the gunfire announced to all that refugees now risked death if they tried to flee East Berlin. On Thursday the East Germans proved their deadly intent. When another man tried to swim across, the guards ran out on a railroad bridge and fired repeatedly down at him until he disappeared underwater. West German frogmen recovered his body three hours later.[8]

Because the Soviets had restricted their activities to their own zone, any action by the West to interfere could have been seen as aggression, triggering greater violence. Despite the apparent injustice to the East Germans, tearing down the wall by force wasn't worth risking nuclear war.

West Berlin Mayor Willy Brandt noted with disgust that the West's inaction would cause "the entire East . . . to laugh from Pankow to Vladivostok."[9]


## KENNEDY


For President John F. Kennedy — a major part of whose presidential campaign was an aggressive anticommunist stance — the Berlin Wall was only one in a string of humiliations. Eight months earlier, for instance, the CIA-led attempt to invade Cuba and overthrow Castro had ended in total failure. When Kennedy refused to lend direct military support to the Bay of Pigs invasion, the 1,200 man rebel force was quickly overcome.[10] "How could I have been so stupid as to let them go ahead?" Kennedy complained privately to his advisors.[11]

In the race to dominate space, things were going badly as well. The National Aeronautics and Space Administration (NASA) had announced the United States' intention to put the first man into space sometime in the spring of 1961. The agency hoped that this flight would prove that the leader of the capitalist world still dominated the fields of technology, science, and exploration.

Originally scheduled for a March 6, 1961 launch, the short fifteen minute suborbital flight was repeatedly delayed. The Mercury capsule's first test flight in January, with a chimpanzee as test pilot, rose forty miles higher than intended, overshot its landing by a hundred and thirty miles, and when the capsule was recovered three hours later it had begun leaking and was actually sinking. Then in March another test of the Mercury capsule included the premature firing of the escape rocket on top of the capsule, the unplanned release of the backup parachutes during descent, and the discovery of dents on the capsule itself.[12]

These difficulties caused NASA to postpone repeatedly its first manned mission. First the agency rescheduled the launch to late March. Then early April. Then mid-April. And then it was too late.

On April 12th, Tass, the Soviet news agency, proudly announced to the world that Yuri Gagarin had become the first human to enter space. Unlike NASA's planned fifteen minute suborbital flight, Gagarin's launch vehicle had reached escape velocity and orbited the earth. As the *New York Times* noted in an editorial, "The political and psychological importance [of this accomplishment gives] the Soviet Union once again the 'high ground' in world prestige."[13] Or as the Soviet government and the Central Committee of the Communist Party stated, "In this achievement, which will pass into history, are embodied the genius of the Soviet people and the powerful force of socialism."[14]

Three weeks after Gagarin's flight, the United States finally entered the space race. Unlike the Soviet launch, where press coverage had been tightly controlled and no public announcements made until the mission was completed and successful, hundreds of newspapermen swarmed about Cape Canaveral.

Twice this first American space flight was scrubbed due to bad weather. Finally, on May 5th at 10:34 AM (two and a half hours late) the Redstone rocket lifted off, pushing astronaut Alan Shepard to 115 miles in altitude before quickly descending to splashdown 302 miles off the coast of Florida. In all, this first American space flight lasted fifteen minutes and twenty-eight seconds, traveling at most 4,500 miles per hour. Compared to the Soviet achievements, it seemed almost pitiful. Gagarin had traveled a hundred times farther, four times faster, and six times longer. And his rocket had put almost four times the weight, five tons, into orbit.

To President Kennedy, this Soviet superiority simply could not be allowed to stand. On May 25, 1961, three weeks after Shepard's short hop into space, Kennedy stood before Congress to deliver what some dubbed his "second" State of the Union speech. He opened by bluntly saying what he saw as his country's role in the Cold War that was raging across the globe.

> We stand for freedom . . . No friend, no neutral, and no adversary should think otherwise. We are not against any man — or any nation — or any system — except as it is hostile to freedom.[15]

He then outlined a number of proposals for increasing the American effort in this "great battleground for the defense and expansion of freedom." He called for additional funds to finance radio and television broadcasts in South America. He reaffirmed his commitment to NATO, pledging at least five more nuclear submarines to that alliance. He described plans to reorganize the Army (giving it greater flexibility) and to increase the size of the Marine Corps. He called for a renewal of the civil defense program, tripling its funding and the building of more fallout shelters.

And then he asked for something more.

> If we are to win the battle that is going on around the world between freedom and tyranny, if we are to win the battle for men's minds, the dramatic achievements in space which occurred in recent weeks should have made clear to us all, as did the Sputnik in 1957, the impact of this adventure on the minds of men everywhere who are attempting to make a determination of which road they should take. Now it is time to take longer strides — time for a great new American enterprise — time for this nation to take a clearly leading role in space achievement which in many ways may hold the key to our future on earth.

Kennedy had no illusions about his reasons for accepting the Soviet challenge to a space race. After noting to all how the Soviets had a clear "head start" with "their large rocket engines," and also noting that his country was willing to take "the additional risk" of joining that space race in full view of the world, he reiterated the political issues that underlay his decision: "We go into space because whatever mankind must undertake, free men must fully share."

To have found a congressmen or senator who opposed this position would have been difficult that night. Nonetheless, all were stunned to silence when Kennedy made his next proposal.

> I believe that this nation should commit itself to achieving the goal, before this decade is out, of landing a man on the moon and returning him safely to earth. No single space project in this period will be more exciting, or more impressive to mankind, or more important for the long range exploration of space.

Many have forgotten in the ensuing decades that Kennedy did not propose this project merely to prove that America could achieve glorious and bold triumphs. When he made this commitment, Americans were truly frightened by the possibility that the Soviet empire was beginning to outstrip them in technology. Worse, this technology gave the Soviets the ability to launch missiles directly at the United States. Khrushchev's words, "We will bury you!" hung over Congress like a thunderhead. Many, both in and out of Washington, believed that their lives and the future of everything they believed in depended on the success of Kennedy's proposal.[16]

## HOUSTON

On September 17th, 1961, nine men out of a pool of 253 applicants were chosen to become America's second class of astronauts, joining the original seven. The list included Neil Armstrong, Pete Conrad, and John Young, the first, third and ninth men to walk on the moon, as well as Ed White, the first American to walk in space.

Also in that list were Frank Borman and Jim Lovell.

For Lovell, it was a dream come true. When NASA picked the seven Mercury astronauts in 1959, Lovell was among the 110 test pilots screened as possible candidates. For several weeks he endured a series of absurd and painful tests. "The enemas," he remembered. "They did enemas all over the place." The most ridiculous test, however, had the doctors strapping his arm to a table, palm up. They took a large needle with an electric wire attached and pierced this through the heel of his hand, just above his wrist. As the doctors gathered around an oscilloscope to watch, they sent a powerful and painful electric charge through the needle, making Lovell's fist involuntarily ball up.

Unfortunately, the oscilloscope wasn't working, so they removed the needle, called in a television repairman to fix it, and then started over. Several times they sent a charge through Lovell's hand, eyeing the meter as if it were the Delphic Oracle. The patient meanwhile was writhing in pain, his hand balling up and opening and balling up and opening.

In the end, Lovell was rejected because he had a slightly high level of bilirubin in his liver. He wasn't sick, it didn't affect his performance, but the doctors needed some criteria to reduce the list. Out he went.[17]

When three years later NASA announced a second call for new astronaut applicants, Lovell jumped at the chance. He was then flight instructor at the Oceana Naval Air Station in Virginia Beach, teaching pilots how to handle the increasingly complex hardware and weaponry of the modern fighter jet. Though the work was interesting and challenging, it still wasn't rocketry or space exploration. Lovell could only watch from a distance as other military test pilots no more qualified than he flew in space, or tested experimental aircraft like the X-15.

He had made it into the Naval Academy on the second try. Maybe the same thing would happen with NASA. He sent in his application, and was quickly accepted.

Frank Borman meanwhile had decided that being a test pilot was no longer advancing his Air Force career. It was important and exciting work, but he wanted more. He had seen how the Mercury space flights had moved the public and the world. He also sensed that the men who became astronauts had a chance to make history. And like Kennedy, Borman strongly believed that what these astronauts did might very well have a direct influence on the outcome of the Cold War.

When NASA announced its call for new astronaut applicants, Frank asked Susan what she thought. Her answer was unequivocal. "I'll do whatever you think is best for your career." In the twelve years of their marriage, she had endured both cinderblock homes in the burning desert and Quonset huts

Frank and Susan Borman in the Philippines, 1953.

Credit: Borman

in the steaming jungle. She had witnessed plane crashes and seen smoke rise from airfields, knowing that any one of those accidents could have claimed her husband.

And she had even risked a plane crash of her own. When Frank had been assigned to the Philippines in 1951, he secured a government house for them to live in, if she could get there by a certain date. Since an ocean voyage would take too long, Susan would have to fly there with three-month-old Fred.

The Bormans, like most young military couples, had very little money. The only thing of value they owned was their car. She sold it for $1,500, which happened to be exactly enough to pay the airfare for herself and Fred.

Owning nothing but her luggage, she and Fred climbed into a Pan Am clipper for the initial part of the journey, a ten hour flight from Los Angeles to Honolulu. From there they would fly another ten hours to Guam, followed by another seven hour flight to Manila.

Halfway to Hawaii, with almost five hours to go, Susan glanced out her window and noticed that one engine was smoking badly. After a few minutes the propeller stopped, and the captain appeared to explain to all the passengers that though they had lost one engine, the plane still had three left, and would make it to Honolulu.

Thirty minutes later Susan looked outside again and was startled to see a second engine also grind to a halt. The captain appeared a second time. With only two engines left on one wing, it was questionable whether they could stay in the air for another four hours. He warned the twenty or so passengers that they might have to ditch the plane in the Pacific. He and the stewardess began drilling everyone on the exit procedures should they land in the water.

Now everyone's eyes were glued to the engines on the other wing. Soon Susan could see these smoking as well, overheating from the strain.

The crew opened a rear door and began throwing things out to lighten the load. Everything unnecessary was tossed away, from blankets to food to liquor supplies.

Rather than try to reach Honolulu, they aimed for the easternmost Hawaiian city of Hilo. With his last two engines belching fire and smoke, the pilot managed to bring the plane in for a smooth landing. The passengers then made an emergency exit. Fred was strapped to her chest, and Susan was the first to slide down the ramp and into the hands of a crew of airport firemen.

Her airplane troubles weren't over, however. Her next two attempts to leave Hawaii, on replacement Pan Am Clippers, were both forced back because of engine troubles. All told, it took her three more days to get to Manila.

A year and a half later, she gave birth to Ed, her second-born. "It was just like the TV show *M*A*S*H*." She was in one hut when the labor pains began, and they had to wheel her across the camp to the operation room, with expectant father Frank trailing alongside.

And yet, Susan wouldn't have traded these tribulations for the world. She and Frank loved each other and had two growing children — what matter they were poor and sometimes endured risk? They were a family and would fight it out, together.

Now Frank wanted to become an astronaut. Susan figured that it couldn't be much different from being a test pilot, with immeasurably better possibilities for success. She was with Frank all the way.

The Bormans arrived in Houston in the late fall of 1962, checking into the Rice Hotel as per orders. The Manned Spacecraft Center was being built in a swampy empty field forty miles south of the city at Clear Lake, and until the astronauts built their own homes, they would live in Houston and commute.

The men went off to the Manned Spacecraft Center to build rockets, fly simulators, and learn everything they could about the space capsules that would send them into space. The hours were brutal, the work was intense and never-ending, and the challenge exhilarating.

The women were left with the job of building a community in the empty farm fields near Clear Lake. The Bormans hadn't even completely unpacked when Frank told Susan that it was up to her to find some land near the Space Center and have a house built. Then he left for Florida, as ordered, to witness the launch of Wally Schirra in the fifth Mercury space flight.

"I had been a pampered and innocent child," she remembers. They had always lived in government housing, had never even rented an apartment. Now she had to create a home from scratch, on her own, on an astronaut's salary of about ten thousand dollars per year. It was exciting, and frightening.

She and Faye Stafford, wife of astronaut Tom Stafford, teamed up and drove down to Clear Lake together. Nothing was there, neither homes nor schools nor shopping centers. "All we saw were cows and fields." They found a real estate agent, purchased some land, and started construction. A dam was built and the swamp drained, and on the shores of this newly created lake the development of El Lago was born.

Meanwhile, Marilyn Lovell was commuting back and forth from Virginia Beach to Houston, selling a home at one place while supervising construction of a new one at the other. The Houston house would be on the north shore of the same lake, in a small development called Timber Cove.

Like Susan, she had backed her husband all the way when he decided to become an astronaut. Marilyn knew that she could have done little to change his mind. Nor did she wish to. She remembered how, when they both were teenagers, he would take her up to the rooftop of his apartment building and show her the constellations and stars. His eyes would shine as he gazed at the stars and talked of going there.

Credit: Lovell

1953. Blanch Lovell holding Barbara, flanked by Marilyn and Jim.

By this time the Lovells had three children, Barbara, nine, Jay, seven, and Susan, four. Marilyn went to inspect the local rural school that they would attend, finding a old brick school with bare planks for a floor. The children wore cutoff jeans and were barefoot. The teacher had a paddle hanging on the wall for discipline. "I was mortified," she remembered. Though Marilyn sent her children there at first, she — along with the wave of new settlers from NASA, including scientists, astronauts, engineers, and designers — quickly brought change to this roughhewn setting. Soon a new public school was built, with modern facilities.

Next Marilyn started looking for a church to join. Though Jim had been raised Presbyterian and Marilyn was Lutheran, several of her new neighbors told her about a little country Episcopal church about ten miles to the north in a small town called La Porte.

St. John's Church was a small brick building located on the town's main street. Its priest, Donald M. Raish, had spent most of the nearly fifty years of his life working in the Episcopal Church of the American Southwest. He had just become rector at St. John's.

The Lovells liked both this church and its soft-spoken rector, and quickly became church members.

The Bormans meanwhile were still living in a rented house in Houston, waiting for their new house in El Lago to be finished. It was Christmas 1962, and they felt isolated living in a city. Both preferred the close-knit community of a military base.

Each Sunday evening they would go to a nearby Houston Episcopal church for services. One Sunday another couple came up to them after church. "We know who *you* are!" the woman said with a grin. Jim and Margaret Elkins lived in Houston with their own three children, and had noticed how alone the Bormans seemed. They invited Frank and Susan to their home to have Christmas dinner. Very quickly the two couples hit it off, becoming close friends. They spent almost every weekend together, going to high school football games or to the Elkins's lake house north of Houston to get away.

At the same time, Susan Borman joined several other astronaut wives to create what they called an astronaut wives' club. Even if they no longer lived on a military base with its community and traditions they could still recreate these in a civilian setting. They published a mimeographed newsletter and scheduled group activities for themselves and their families.

Soon what had been cowfields in the Texas countryside was a thriving American town, no different from a hundred thousand other communities across the nation.

\* \* \*

While the Bormans and Lovells were settling into the more comfortable but hectic public life of an astronaut's family, Bill Anders was still struggling to get into the test pilot program. The Anders were now at Wright-Patterson Air Force Base in Dayton, Ohio where Bill attended the Air Force Institute of Technology.

Not only did Bill work like mad taking nuclear engineering, he went to night school at Ohio State, studying aircraft stability and control. He also rose early each morning and went flying for several hours before breakfast, just to log more flight time.

They now had three children, Alan, five, Glen, four, and Gayle, one, and were living in their first home, a four bedroom brick "palace" purchased for under $15,000.

Each night Bill and a number of the other student officers would gather in his study. Valerie would bring them food and coffee, and for hours they would review physics and engineering problems.

Bill's schedule was so tight that the only break he took was fifteen minutes each evening to eat dinner and watch the news. Sometimes one-year-old Gayle would crawl to Bill's study door, lie on her belly and put her hands through the gap at the base of the door. Bill would see her fingers, come out to hold her for a while, then go back inside.

For Valerie, the price of marrying a man who wanted to be an astronaut was as hard and unrelenting as her husband's schedule. She, like Marilyn and Susan, tolerated miserable living conditions in the 1950's in order to further her husband's career. At Bill's first assignment in the Rio Grande Valley of Texas, the couple lived for a time in a tiny one-room storefront office, with no air conditioning and the only bedroom window made of glass bricks that couldn't open. This was just as well, as the bedroom faced the town's main street and "there was a cricket epidemic," Valerie remembered. "The whole town was filled with black crickets everywhere." Valerie, used to the comfortable San Diego climate, soon developed a kidney infection which was followed by infectious mononucleosis.

Nor were the sacrifices she and the other wives faced merely discomfort and  meager living conditions. They also sacrificed careers and university degrees. All three women attended college — something few were able to do in the 1950's. Wherever Valerie lived she enrolled at the nearest university, taking courses from astronomy to oceanography, simply because the subjects interested her. Later, when NASA did a background check on her husband to see if he was qualified to become an astronaut, it also investigated Valerie, making sure that she, like Marilyn and Susan, could handle the pressures and challenges she was certain to face. Just as he had to measure up, so did she.

February, 1959. Valerie Anders lived in her grandmother's home in San Diego while Bill was stationed in Iceland. She holds Alan, with her grandmother Babette Prasser on the left and her mother Elsie Hoard on the right, holding Glen.

The women accepted these sacrifices for the sake of their husbands and children. They knew that the men they loved were unique, destined to achieve great things. And they knew that if they did their part, their respective spouse's achievements might even be greater, for themselves and their nation.

For the Anderses, the sacrifices and hard work paid off. Bill graduated from the Institute of Technology with high honors and, having finally obtained the advanced degree that the test pilots' school at Edwards demanded, he went back to reapply.

In the interim two years, however, the test pilot school had changed the rules. Now they didn't want schooling, they wanted pilots with lots of flying experience. Had he not gone to school the last two years he would have been a shoo-in. Once again he was locked out.

Nonetheless, he kept applying. The family moved to Kirkland Air Force Base in Albuquerque, New Mexico where Bill became an engineer and flight instructor. There, he and a friend periodically took a plane and flew to Edwards to schmooze and play politics. Sometimes he flew demonstrations, even repeating the flying maneuvers required by the entrance exam.

He might not have the flying time they wanted, but he could do anything in the air that anyone else could do. Bill Anders was determined not to take no for an answer.


## BERLIN

On August 13th, 1962, the first anniversary of the construction of the Berlin Wall passed. Several thousand demonstrators gathered along the wall, throwing bottles and stones at the East German guards. At one point the crowd threw a paving stone through the window of the Soviet Intourist travel agency, located near the wall in West Berlin. The East Germans in turn responded with tear gas and water cannon.

Few in America noticed. Two days earlier, on August 11th, the United States had been shocked by yet another Soviet first in space. In less than twenty-four hours the Soviets had launched their third and fourth cosmonauts into space. For three days Andrian Nikolayev and Pavel Popovich orbited the earth, talking and actually singing duets as their separate capsules passed each other in space.

Once again American commentators were appalled at the seemingly insurmountable Soviet lead in space. Not only had the communists launched two rockets in quick succession, they kept two men in space for just under three and four days respectively, and their craft actually seemed to demonstrate an ability to perform a rendezvous in space.* Near the beginning of their mission the spacecraft were only four miles apart, and the cosmonauts reported they were close enough to see each other's capsule.

---

* Though Soviet newspapers made this claim, neither craft had the ability to maneuver in space, and were never piloted towards each other.

A few commentators wondered whether Khrushchev had deliberately timed the two space shots to occur during the Berlin Wall's first anniversary. Whatever his intentions, the space flights certainly dominated world news. With the safe return of the cosmonauts on August 15th the Soviet government triumphed their success, noting that "communism is scoring one victory after another in its peaceful competition with capitalism."[18] The *New York Times*, meanwhile, editorialized that "The Jules Verne journey of the two Soviet cosmonauts now safely back from their rendezvous in space. . .is a spectacular accomplishment, an amazing feat."[19]

While cosmonauts Nikolivich and Popovich were being lauded and celebrated in Moscow, two East Berliners decided they would make their own statement about the merits of communism.

Peter Fechter, 18, a construction worker in East Berlin, had had enough of living behind the wall. As he wrote to his sister (who had fled to West Berlin in 1956), "Those swine have stepped up our work quotas again so we lose fifty to sixty pfennigs an hour and have to work ten hours to earn what we used to make in 8 1/2."[20]

On August 17th, four days after the wall's first anniversary and two days after the return of the cosmonauts, he and co-worker Helmet Kulbeik broke for lunch, a meal of boiled bacon and potatoes. It was the first time he had had bacon in four months.

Then the two men walked over to a part of the wall very close to Checkpoint Charlie, the one remaining border crossing between West and East Berlin. Two days earlier they had scouted out the area and discovered a construction crew hard at work renovating an abandoned building that abutted the wall. Because the two youths also wore construction clothes, no one had questioned them as they entered the building and wandered from floor to floor. To their surprise, one ground floor window facing the Berlin Wall was not bricked up but was instead blocked only with barbed wire and wooden planks.

Now they entered again and went straight to that window. They ripped out the boards and pulled the barbed wire clear.

Before them was the hundred foot-wide barricade of barbed wire rolls leading to the eight-foot-high wire fence of the wall. Patrolling that death strip were machine gun-carrying East German soldiers, who in the last year had killed forty-nine people trying to escape East Berlin.

After what was almost certainly a long moment's hesitation, both men made a sudden dash for freedom, tearing and stumbling their way through the jungle of barbed wire scattered across that deadman's zone. Kulbeik managed to reach the wall first. He leaped upon the wire and climbed. Fechter was right behind him. On the East German side of the deadman's zone, border guards unshouldered their rifles and screamed for the two men to halt. As Kulbeik pulled himself over the four strands of barbed wire at the wall's top the guards opened fire. Fechter, who was halfway up the wall, was hit in the back and stomach. He screamed in pain, but held on. Kulbeik reached down to try and help him up, and for a second the two struggled to get Fechter over the top.

Then Fechter fell back into the death strip. Unable to do anything to help him, Kulbeik jumped down into West Berlin and to freedom.

Fechter lay there at the wall's base, slowly bleeding to death. Though he was less than two feet from the American zone and in plain sight through the wires, there was nothing anyone could do. When several West Berliners started to climb the wire to help him the East German guards threw tear gas to drive them back. And American soldiers could only throw Fechter some bandages: they were forbidden to cross the wall into East German territory.

On the East German side, the guards were afraid to come out and get him. He had fallen so close to the wall that they feared attack from the growing West German crowd just on the other side of the wire barrier.

For almost an hour Fechter lay there, groaning in pain. After awhile his groans stopped.

Finally, with heavy reinforcements covering them, the East German guards came out and carried Fechter's body away — the fiftieth person killed trying to breach the Berlin Wall. Within hours thousands gathered at the wall, and as they had four days earlier, they threw rocks and bottles at the East German guards.[21]

Though Fechter's death made front page news in America, it failed entirely to distract the world from the just-completed Soviet space triumph.[22] Standing in Red Square before a huge crowd of citizens, Khrushchev and others repeatedly proclaimed communism's supremacy. As cosmonaut Nikolayev noted to the crowd, "The group flight in outer space is one more vivid proof of the superiority of socialism over capitalism."[23]

Few could argue. The day before, D. Brainerd Holmes, director of NASA's manned space programs at the time, told the press that the launch of the next U.S. manned flight would likely be delayed. Furthermore, he admitted that it would be years before the U.S. could launch two astronauts into space at the same time, simply because the U.S. only had one launchpad.[24]

Thirteen years had passed since the Berlin airlift. Five years had passed since the dawn of the space race. One year had passed since the construction of the Berlin Wall. Despite the efforts of many in the West, it seemed that cosmonaut Andrian Nikolayev might very well be right.

## CHAPTER FIVE

# "WELCOME TO THE MOON'S SPHERE."

EIGHT AM (C.S.T.), MONDAY MORNING, DECEMBER 23, 1968. In the Apollo 8 capsule Jim Lovell was fast asleep, Frank Borman had just gotten up, and Bill Anders was at the controls. As he had on Sunday, Mike Collins started his shift with what he now called "the 23rd of December edition of the *Interstellar Times*," a quick summary of some of the more interesting news items of the last few hours.

He began by warning the astronauts that "there are only two more shopping days until Christmas," then described how twenty-three convicts had escaped from a New Orleans prison, how President Johnson had sent the astronauts his congratulations, how a big blizzard had hit the Midwest, and how the football playoffs were shaping up.

Borman asked, "How are the families doing, Mike?"

This was not the kind of question that Frank Borman would usually ask during a mission. And since his voice and Bill Anders's sounded were very much alike, thin altos compared to Lovell's rich bass, Collins assumed Anders had asked the question. "They are doing just great, Bill; just talked to Valerie a few minutes ago." He had called her from home, just before leaving for his shift.

"That was Frank," Borman said.

"Oh, well, likewise with Susan," Collins recovered. "I have not talked to her since last night."

"Roger."

For Susan Borman, the battle was not so much over fighting the worry and fear, but preventing anyone from finding out how afraid she was. Her solution was to dull her mind. She would mix herself a drink and try and play hostess as neighbors and friends arrived with their encouragement and food.

Helping her were her two teenage sons. With the fearlessness of youth and the same boundless confidence of their father, both boys were heedless of the dangers. Separated from his daily grind and high-pressure concerns, they didn't understand the risk and were instead sure that everything would work out. This was simply his day job, and he enjoyed doing it.

Their mother, however, was always close by, and they could see how obsessed and worried she was about the flight, almost to the exclusion of food and sleep. In fact, she had eaten so little since launch that at one point Fred sat down with her and demanded that Susan eat something. She shook her head. Food was the last thing on her mind.

Undeterred, he put potato salad on a fork and thrust it at her. "Eat!" he insisted. When she still refused, he began to imitate how she would treat him when he would refuse food as a baby. "Open the hanger door, here comes the plane," he sang, aiming the fork at her mouth like a airplane. "RrrrRRRrrRRRRR," he rumbled, simulating the sound of a propeller plane.

Susan laughed, and took a mouthful.[1]

* * *

Just as she had on Saturday night, Marilyn Lovell had difficulty sleeping Sunday. She had dozed, much like the astronauts, sleeping in short restless bursts. Periodically she would get up, go into the kitchen to listen to their voices on the squawk box while smoking another cigarette.

Dawn finally arrived, and as scheduled she went to her Monday morning beauty parlor appointment. Then she did some shopping at the local grocery store. At some point that morning she went to visit the Borman and Anders homes.

Susan Borman flanked by Fred on left and Ed on right,
taken during the flight. Credit: Borman

Her kids, meanwhile, were doing what kids normally do. While fifteen-year-old Barbara went shopping with her high school friends and ten-year-old Susan spent time playing with friends in Timber Cove, thirteen-year-old Jay was having stomach problems. In order to get him to the doctors at NASA without the press noticing, she slipped him out the back door and hid him in the back seat of her next-door neighbor's car with a blanket over his head. At NASA the doctors told her that the boy was merely upset because of all the excitement.

Meanwhile, two-year-old Jeffrey periodically opened the front door and held his own impromptu press conferences with the reporters stationed there. On his head he wore his own little astronaut helmet, which he proudly showed off to the press.

By midday the Lovell house once again began filling with people, most of whom were women who like Marilyn attended St. John's church. Father Raish also came by. At one point he suggested that they hold communion right there, since Marilyn hadn't been able to get to church that weekend.

"Father Raish was a very warm human being, and he sensed when he was needed," Marilyn remembered. In the six years since the family had moved to Houston, he had become a special person to her. Because of Jim's heavy work schedule she was often alone, and he frequently made it a point to stop by the house in the late afternoon. They would sit and chat. "I could bare my soul to him," she said years later.

The women immediately agreed to Father Raish's suggestion. Together they knelt around the family room coffee table, and he led them in prayer.

\* \* \*

Valerie Anders had gotten up early Monday morning to dress and feed the kids. She found that her youngest ones, Eric, four, and Greg, six, had become more fussy and needy, while Gayle, eight, had started to suck her thumb again for the first time in months.

The mob of reporters were still on her lawn, trapping her in her home. Since the day was cold she opened her garage and put a large pot of coffee there for them.

\* \* \*

On the spacecraft things continued to go well. One minor problem, a chilly cabin temperature, had been solved during the night by turning on all the cabin fans (which the astronauts had shut down because one in particular was very noisy).

The astronauts were now over 188,000 miles from earth. Their radio signal, moving at the speed of light, took more than a full second to get home.

And yet, even after two days of travel, they still had almost twenty more hours before they would reach the moon. The waiting continued.

Anders found himself both bored and edgy. After more than forty-eight hours in space, he had officially rested only six hours, the least of all three men, and had actually slept much less. Even taking a sleeping pill on Sunday afternoon had not helped. He found that the combined excitement and tension of his first space flight would not let him relax. Nor did it help that, as much as they tried to keep quiet, Lovell and Borman liked to talk.

Yet, when he wasn't trying to sleep he was startled by how surprisingly tedious space flight was. He sat, scanning the dials again and again for problems, constantly updating himself on the spacecraft's status. Everything was running perfectly. Periodically he did some basic maintenance chore, such as purging the fuel cell batteries to keep them running, or switching antennas as the spacecraft rotated.

And he stared out the window, finding that the only thing he had to look at was a steadily shrinking earth, drifting across his window once a minute as the spacecraft gently rotated. The moon he had not seen. With the spacecraft pointed tail-first at the moon, the windows never faced it. As he later said to mission control, "It's like being on the inside of a submarine."

* * *

At 10:30 AM Valerie Anders decided to make a break from her home and visit mission control. Leaving the kids in the care of au pair Silvie, she was escorted by NASA press officials through the gauntlet of reporters to a NASA car and driven to the Manned Spacecraft Center.

Once there she went to the private lounge positioned behind communications. She waved to Mike Collins, and sat down to watch for a while. Nearby sat George Low, manager of the Apollo Spacecraft Program. Low had taken over the program in January 1967 following the launchpad fire that had killed Gus Grissom, Ed White, and Roger Chaffee. Less than eighteen months later, he became the man who pushed NASA to send Apollo 8 to the moon.

He and Valerie chatted. She quickly saw that, though he tried to hide it, seeing her made him very nervous. If something went wrong he would have brought disaster to her and her five little children. His concern touched her, and Valerie tried to ease his mind by being bright and cheery and talking about how well everything was going. She had faith in George Low, and she wanted to show him that.

After a few minutes she heard Bill tell Mike Collins that he was passing the controls to Borman so that he could "take a little snooze for a while." Quickly Valerie passed a message to Collins. She wanted him to tell Frank to tell Bill that she wished him "a happy nap." In the spacecraft, Borman grinned at Anders, now floating in the small space below the couches and trying (but failing) to sleep. "Okay," he said to Collins. "Tell her that he makes us tired sometimes too."

About this time Jim Lovell finally woke up, having slept for almost seven hours. As the mission's navigator, he now began a long discussion with the ground on what stars to look at and where he should sight. He reiterated his problems with fuel and light glare, noting that unless he allowed his eyes to completely adapt to the darkness of space he could not see the stars.

Despite these difficulties, however, on Sunday Lovell had managed to make his own estimate of the spacecraft's course and position. When Houston compared calculations, they found that his numbers were "within a couple of a thousandths of a degree of the theoretical optimum." Mattingly had added jokingly, "Well, I am getting a lot of confidence in your ability to run that mystery show now."

Anders had responded, "Hey, we have to spend four more days up here with him, will you take it easy? [Jim] is already talking about going back to M.I.T. as a professor."

The ground had discovered another interesting navigational phenomenon. When the astronauts dumped their waste urine overboard, it actually acted as a propellant and changed their course slightly. This unexpected power source eventually required them to do a small additional mid-course correction.

Time passed. The astronauts dozed, or did other routine maintenance.

* * *

Finally, at 12:30 PM, Ken Mattingly, now at communications, asked Frank Borman about their second planned television show. "Are you planning to show us TV pictures of the earth today?"

"Well, that is what we wanted to do," Borman answered. "It seems that would be the most interesting thing we can show you, but we, you know, had trouble with the lens."

Mattingly then began describing to Borman the solution proposed by the video technicians on the ground. First the astronauts needed to take one of the filters from the film cameras and duct tape it to the front of the video lens. Then, with the camera mounted in its bracket, the men could aim the camera by reorienting the spacecraft. "Do not touch the body of the lens while televising," Mattingly told Borman. "Apparently if you put your hands on the [telephoto] lens itself, it causes electrical interference." They were also warned to give the lens's automatic light meter from ten to twenty seconds to warm up. The technicians suspected that the previous day the lens hadn't had time to adjust to the high contrast of light coming from a bright earth surrounded by a black sky. For more than an hour Mattingly and Borman went over these steps, with Mattingly noting that "the show as scheduled is just out the window at the earth only."

Borman agreed, though he had his doubts about the lens. "I bet the TV doesn't work."

Mattingly hedged, "Well, we won't take that bet, but anyway, we are standing by for a nice lurid description [of the earth]."

At 2 PM, they turned on the camera. With Borman as pilot and Anders as cameraman, Lovell became the narrator. Because the camera had no eyepiece, the astronauts could only aim it using instructions from earth or, as Anders noted, by "looking down the side or putting some chewing gum on top." Borman and Anders struggled to keep the earth centered in the camera frame, with Borman maneuvering the capsule to make the major adjustments and Anders fine-tuning the picture by tweaking the camera's mounting.

This time the camera lens worked. The astronauts successfully transmitted to earth the first live televised pictures of the home planet as a globe.

In many ways, this telecast foreshadowed today's news coverage, where every major event is televised live, and every citizen can watch it happen merely by pressing a button. Yet, because this type of newscast was unprecedented, there were no announcers, no talking heads "analyzing" what everyone was watching. The moment had a freshness and impact gone from much of modern news broadcasting.

By now Valerie had returned from mission control to find an almost party-like atmosphere at her home. Many of Valerie's friends had come by, and she, her children, Bill's aunt and uncle from Texas, and others crowded around her new color television to watch the telecast. Not surprisingly, Valerie found that her children were more interested in watching cartoons or going outside to play with friends. However, she insisted that they stay and watch.

Susan Borman was also pinned to the television. But her two sons had had enough of the crowds of people and the mob of reporters. That morning they had both decided to go duck hunting in an effort to get away from all the hoopla. They got into their hunting coveralls and loaded their gear into Fred's car, planning to drive out to the country farm of a family friend.

The photographer from *Life* magazine noticed what they were doing and decided it would make a great picture to see them arm-in-arm with their mother holding their shotguns. Susan hated these posed shots, but she went outside in her backyard to do her job. The boys simply wanted to get away as quickly as possible, so they obliged as well.

Then they climbed into Fred's car and headed out. Not surprisingly, a handful of journalists jumped into cars to follow, and for a few minutes the boys led the world's press corps on a merry chase through El Lago and Timber Cove. To get away Fred turned into the front entrance of NASA to cut through the Manned Spaceflight Center and exit its back gate. While the Bormans could pass through quickly, their shadowers were left at the main gate, trying to get clearance. Fred and Ed then disappeared into the country for a few hours' relief from the constant stream of visitors at their home.

At the Lovell house the situation was similar. While most of her family and friends gathered in front of the television, Marilyn almost had to drag Jay in from the backyard. He wasn't interested in seeing a grainy, black-and-white picture of the earth when he could be out with friends on this Christmas vacation day.

On the television Jim Lovell began by noting how the earth's observable size was quite small: "About as big as the end of my thumb" when held at arm's length. He then described the visible continents, from the North Pole on top to the southern tip of South America at the bottom. "For colors, waters are all sort

The earth, December 23, 1968. This televised view matches the 70mm
still photo in the color section.

of a royal blue; clouds, of course, are bright white . . . the land areas are generally
a brownish — sort of dark brownish to light brown in texture."

Everyone, in space as well as earth, could see that there was absolutely
no visible evidence that a civilization of more than three billion people existed
on that small planet.

Lovell in particular was struck by the scale of the cosmos. His
navigation work, sighting off the moon, the earth, the sun, and the stars, had
given him a real sense of where he was in that vastness. And though he had
at that moment spent more time in space than any other human (having

flown two weeks with Borman in 1965 and four days with Buzz Aldrin in 1966), he found himself awed at the smallness of earth in that black sky. As he told his earthbound audience, "What I keep imagining is, if I were a traveler from another planet, what would I think about the earth at this altitude — would I think it was inhabited?"

To Bill Anders, that tiny blue-white earth suddenly reminded him of the Christmas tree ornaments he and Valerie had hung only three weeks before. The planet was round, it glittered, and its surface seemed delicate and easily destroyed.

While they didn't say so, and none of them would have even admitted it to themselves, the three men were truly staggered by the immensity of the emptiness around them, and the jewel-like splendor of the shimmering sphere floating within it.

Nor were they alone in their impression. On earth one could feel it in their words, and see it in the televised image of the earth. As Anders noted during the broadcast, "You are looking at yourselves at [200,000] miles out in space."

Valerie Anders, focusing her mind on the excitement and joy of the space flight, gazed at the television with utter elation. She had trouble imagining how far away the astronauts were. *That's the earth*, she thought incredulously, *with all its billions of people on it.*

Susan Borman looked at this vision of the earth and felt only disbelief. *How could this be?*, she thought. She found herself struggling to imagine Frank actually in space so far from home. To Susan, the television image actually made it all seem less real, like a science fiction television show. So, rather than watch, she closed her eyes and listened to Frank's voice. When she did, she found that she could imagine him in the capsule, surrounded by infinite space.

It wasn't enough. She couldn't put herself there with him. Nor could her fantasizing help get him home.

For Marilyn Lovell, the vastness of space and the danger that surrounded her husband finally hit her. Jim unwittingly helped pound the point home when, still amazed at how little of human civilization he could identify from this distance, he wondered aloud if a visitor from the stars would know to "land on the blue or the brown part of the earth."

Bill Anders joked, "You better hope that we land on the blue part."

As the telecast ended, Marilyn's children scattered, and suddenly Marilyn felt a compelling need to get closer to her husband. She decided to make a quick trip to mission control to watch the action and reassure herself that all would be fine. The NASA liaison officer quickly arranged for a ride, and in less than ten minutes she was in the V.I.P. section of the control room.

Her timing couldn't have been worse. Almost as soon as she arrived the Apollo 8 spacecraft passed the point in which the moon's and earth's gravity balanced. Up until that moment the earth had been pulling at them, slowing them down until their ship's speed had dropped to only 2,223 miles per hour. Now they had crossed into what the engineers called "the moon's sphere of influence," its gravity pulling at them and drawing them in. Jerry Carr, at capcom, made a point of letting Borman know. "By the way, welcome to the moon's sphere."

At first Borman didn't understand. "The moon's fair?" he asked, puzzled.

"The moon's sphere," Carr said, more slowly. "You're in the influence."

Borman joked, "That's better than being under the influence."

Marilyn left less reassured than when she arrived, and returned to a strangely quiet house. Her kids were off playing somewhere, and the crowds of friends had drifted away when she had gone to mission control. Marilyn was completely alone for the first time since the launch.

She stood in her empty home, listening to the hollow sound of the squawk box quietly hissing its never-ending stream of technical jargon.

*Carr: Apollo 8, Houston. We're dumping at this time.*

*Anders: Roger. Tape voice is probable. We ought to get a check on it as low bit rate for D.S.E. voice.*

*Carr: Apollo 8, are you saying that everything that's on there now is in high bit?*

*Anders: That's where my switch was.*

*Carr: Okay. We'll take a look at it then . . .*

Then Paul Haney, the public voice of NASA, noted that the spacecraft's speed was increasing. The capsule was now more than 205,000 miles from earth, traveling at over 2,700 miles per hour.

Unexpectedly, Marilyn grasped the reality of her husband's situation. Jim would go into lunar orbit, something would fail, and she would never see him again. She sat down at the little bar between the kitchen and the family room, poured herself a drink, and broke down in tears. For several minutes all she could do was cry, the tension finally breaking in long sobbing gasps.

Not long after, there was a knock on the door. Marilyn wiped her face, took a breath, and went to answer it. There stood young Betsy Benware, the teenage daughter of Betty Benware, one of Marilyn's neighbors. She was holding a tray with a dinner her mother had cooked for Marilyn.

"Are you all right, Mrs. Lovell?" Betsy asked.

"Oh, yes," said Marilyn. She didn't want anyone to worry about her. She took the food appreciatively and sent the girl home.

Within minutes Betty Benware arrived at the Lovell home. She had already called some of their other friends, and soon the house was once again filled with people, gathered there to stay with Marilyn through the night and through the coming lunar visit.

They fed her, and then convinced her to try and get some rest. She went to her bedroom, lay down, and for the first time since the launch fell into a deep sleep. The crying fit had released much of her pent-up tension. She slept for almost seven hours.

* * *

More than two hundred thousand miles away, Apollo 8 flew on, its speed increasing rapidly and irrevocably. It was falling towards the moon, and nothing in the universe could prevent it from getting there.

## CHAPTER SIX

# HUGGING
# THE COAST

KHRUSHCHEV

COSMONAUT ALEXEI A. LEONOV GRASPED THE HATCH and opened it carefully. Outside, the jet black sky surrounded him like a velvet hood. Below rolled the glowing horizon of the world, speeding past at over 17,000 miles per hour.

With a grunt Leonov pulled himself through the hatch, pushing off from the Voskhod spacecraft to slowly drift fifteen feet away. He was floating more than one hundred twenty miles above the earth's surface, his only link to the space capsule a thin twenty-foot tether. "I didn't experience fear. There was only a sense of the infinite expanse and depth of the universe."[1]

It was March 18, 1965, and Leonov had become the first human being to walk in space.

For ten minutes he pirouetted about, waving and smiling at Pavel Belyaev, his commander who was watching from inside Voskhod. Below him the Black Sea rolled by, followed by the Ural Mountains of Russia. Then Belyaev told him that with only forty-five minutes of oxygen left, it was time to come back inside.

Only now Leonov had trouble squeezing himself back through the hatch. The camera that had filmed his adventure kept getting in his way, and his spacesuit had swelled when its internal air pressure pushed against the vacuum of space. For eight minutes he struggled, pushing and pushing again and again in a vain attempt to force his body through the hatch. Finally, with his oxygen supply quickly disappearing, he took a desperate chance and partly depressurized his spacesuit. The release reduced the size of the suit enough so that he could slide in, slamming the hatch behind him.[2]

Once again the Soviets had struck first, beating the Americans in space. "The so-called system of free enterprise is turning out to be powerless in competition with socialism in such a complex and modern area as space research,"[3] proclaimed an article in Pravda. Not only did this first human spacewalk take place just five days before the first launch of the American Gemini space program, the Soviets proudly trumpeted the second success of what they called their new Voskhod spacecraft.

Khrushchev's daring, boisterous, and relentless style of leadership demanded these increasingly dangerous space stunts. For Khrushchev, the Soviet Union had to stay ahead, no matter what. Knowing that in December 1957 the first American satellite would launch, he had pushed for a launch of Sputnik 1 in October. Knowing that Alan Shepard's mission was scheduled for May 1961, he had Yuri Gagarin sent up in April. Knowing that the Americans planned to attempt a rendezvous in space, he had Nikolayev and Popovich launched on their group flight in August 1962.

In 1964 he demanded that his space engineers accomplish two more stunts to beat the Americans. They were to fly a three man space capsule, and have a cosmonaut leave the capsule to walk in space.

More than Leonov's spacewalk, the October 1964 three-man flight of Voskhod 1 epitomized the demands that Nikita Khrushchev put on the Soviet space program.

To do what their ruler demanded, the Soviet engineers took incredible risks. Using the same tiny Vostok capsule that had put Yuri Gagarin in space, they removed its ejector seats and escape tower. Then they eliminated spacesuits, and had the crew sit sideways to the control panel. Then they put this all on top of a brand new rocket that had been tested just once.[4]

The flight was so risky that one of the ship's designers, Konstantin Feoktistov, insisted on flying himself. He said he couldn't ask others to go if he wasn't willing to go himself.[5]

On October 12th, 1964, five months before Leonov's spacewalk and the first American two-man mission, Voskhod 1 took off from Baikonur. For a little over one day designer Feoktistov, test pilot Vladimir Komarov, and doctor Boris Yegorov orbited the earth in their cramped quarters, once again proclaiming to the world that communism under Khrushchev could do it better. In fact, Khrushchev spoke with the cosmonauts while they were in orbit, wishing them health and telling them that their work would "glorify our homeland, our peoples, our party, and the idea of Marxism-Leninism [by] which our state stands and [by] which we achieve all the things we have."[6]

When Voskhod landed on October 13th, however, the Soviet Union was no longer under Khrushchev's rule. In a sudden coup, a political faction led by Leonid Brezhnev had taken control of the government. Khrushchev's freewheeling style had finally done him in. Without mentioning his name, a Pravda editorial condemned Khrushchev's "hare-brained schemes, immature conclusions and hasty decisions and actions divorced from reality, bragging and phrase-mongering, commmandism, [and] unwillingness to take into account the achievements of science."[7]

His continuous interference with the space program in order to achieve short-term propaganda victories had contributed significantly to his ouster.[8] Unbeknownst to NASA and the rest of the world, the new Soviet leadership had decided that, after Leonov's spacewalk five months hence, manned flights would stop for a few years in order to give the space program time to refocus. No longer would its missions be planned merely as solitary stunts to upstage the West. Now the Soviets were going to establish a carefully thought-out program for beating America to the moon.

## GEMINI

Imagine you and a co-worker sit in the front seat of a small compact car. You close the doors and proceed to live in that confined space non-stop for the next fourteen days. You cannot leave to go to the bathroom, to eat, or to shower. Imagine that you have radio headsets on and that every word you say is being recorded by an army of doctors. Imagine also that those doctors have attached sensors to numerous places on your body. They have many different questions to ask you, and you have no choice but to try to answer them.

Imagine as well that the car's air conditioning doesn't work very well, the car is sitting in the hot sun, and you have to wear a heavy, insulated jumpsuit. The temperature rises and there is nothing you can do.

And finally, imagine that though the car's engine is in gear and running and the car is in motion, the steering column is turned all the way to the right, and for the entire two weeks you continually go around in circles, watching the same scenery go by again and again and again, twenty-four hours a day, seven days a week.

This is the experience that began for Frank Borman and Jim Lovell on December 4, 1965 at 1:30 PM (C.S.T.). At that moment they lay on their backs on top of a one-hundred-twenty-foot-tall raging behemoth. The Titan rocket on which their Gemini capsule sat had just ignited, and though it was only a third as tall as the Saturn 5, it was a significantly rougher ride. Spewing out 430,000 pounds of thrust, twice as much as a Boeing 747 at takeoff, the Titan felt like a bucking bull at a rodeo.[9]

It was the first space flight for both men, and in as many ways as possible Gemini 7 illustrated the unpleasant and miserable side to human exploration. Their mission was to prove that a human being could survive fourteen days in space, and for two weeks they went around and around and around and around the earth, completing two hundred six orbits and seeing as many sunrises and sunsets.

For the first two days of Gemini 7, the rules required that Borman and Lovell stay in their spacesuits, which they found hot and uncomfortable. After that, if one astronaut was in shirtsleeves, the other had to be in his suit. The original plan called for them to switch places each day, with Jim Lovell in his longjohns on the third day, and Frank Borman out of the suit on the fourth, and so on.[10] As it turned out, Lovell asked if he could stay unsuited on the second night, and Commander Borman made the decision that since his crewmate was a larger man and had greater difficulty getting out of the

Gemini 7 lifts off, December 4, 1965.

suit, he would stay suited and let Lovell remain in his underwear. "I didn't have the heart to follow the twenty-four hour exchange arrangement," Borman wrote later.[11]

For the next four days Borman sweated in his suit, resisting mission control's repeated requests that he trade places with Lovell so that the doctors could get better data. Instead, Borman argued that there was no reason for either man to wear his suits, and that they should both be allowed to fly suitless.

Because this was the first long duration flight in space, almost doubling the previous mission length, the medical experiments took precedent over the comfort of the astronauts. On the mission's sixth day, with the capsule's internal temperature now at 85 degrees, Flight Director Chris Kraft ordered Borman to switch places with Lovell. In his next rest period Borman slept for six straight hours, and woke up telling ground control that he "felt like a million dollars!"[12]

There were other unpleasant aspects of their mission. The astronauts urinated into condoms which were sealed in a plastic bag and then dumped overboard. On the fifth day the plastic bag broke in Borman's hands, and little globs of urine floated all over the capsule.[13]

Both men found their noses stuffed and their skin flaking due to the one hundred percent oxygen atmosphere. In order to simplify the capsule's design, NASA engineers had eschewed recreating the earth's normal atmosphere of about three-quarters nitrogen and one quarter oxygen. Instead, the Gemini spacecraft used oxygen alone, at a pressure of 5.5 pounds per square inch. A mixed atmosphere required additional pumps, tanks, and valves, weighed more, and cost money and time to build. Two weeks of astronaut discomfort, however, cost nothing in time, labor, or weight.

The dehydrated food the astronauts ate was at best boring, and at worst horrible. "The worst items were the beef and egg bites," Borman recalled later. "Terribly dry and leaving a bad taste in the mouth and a coating on the tongue."[14]

On the eighth day of the mission, NASA finally attempted to match the Soviet accomplishment of three years earlier: putting two manned spacecraft in orbit at the same time.

Gemini 6, manned by Wally Schirra and Tom Stafford, had originally been scheduled to launch in October. First NASA would launch an unmanned target craft. Then the Gemini capsule would follow it into space, chase it down, and link up. To prove such maneuvers possible was essential for any mission to the moon. Unfortunately, when the target rocket exploded six minutes into its flight, Schirra and Stafford were left without anything to rendezvous with, and so their flight was scrubbed.

To salvage the mission, NASA improvised. Why not combine the two week-long endurance mission of Gemini 7 with Gemini 6's rendezvous goal? Eight days after Borman and Lovell reached orbit, Gemini 6 would blast off from the now-renamed Cape Kennedy and, using the Gemini 7 capsule as their target vehicle, track it down.

So, while Borman and Lovell sweated and squirmed through their first week in orbit, ground crews scrambled to repair and prep the launchpad for Gemini 6's flight. In less than two days, the Titan rocket was prepped and ready to go.

In the early hours of December 12th, Wally Schirra and Tom Stafford entered the capsule. As with every other American launch, hundreds of newsmen gathered at the Cape to report the story. All the television networks preempted normal schedules to show the launch. The tension built as the clock wound down. At T minus ten seconds, Paul Haney, public affairs officer, began counting down the last seconds, then announced that the engines had ignited. On the launchpad smoke billowed out from under the rocket, the roar built — and then stopped.

A strange silence descended on the pad. Gemini 6 sat there, unmoving. Haney announced, "We've got — we've got a shutdown! No liftoff! The engines have shut down!"

In the capsule Commander Wally Schirra sat tensed, his hand holding the emergency ejection system release cord. According to the rules he should now pull it, sending the astronauts flying away from what was over one hundred fifty tons of explosive fuel. Instead, he looked at the spacecraft's internal timer and noted aloud to mission control, "My clock has started."

Gemini flight director Chris Kraft cut in, "No liftoff, no liftoff." Without ordering them to do so, Kraft desperately wanted Schirra to pull the cord before the Titan rocket exploded.[15]

Schirra held firm. Like all the astronauts, he had an amazing instinct for knowing when to abandon his ship. Ejection not only would have destroyed any chance of Gemini 6 ever making orbit, it held the real risk of injuring him and Stafford. If the Titan rocket was still locked on the pad, nothing would happen, and once the ground technicians got the rocket fuel under control they could just reset the clocks and start over.

Schirra's gamble paid off. An hour and a half later the two astronauts climbed from their Gemini capsule. At the same time, ground crews swarmed over the rocket, trying to discover why its engines had cut off.

Meanwhile, Borman and Lovell continued to circle the globe. To pass the time, they sang an old country song over and over again. "Put your sweet lips a little closer to the phone, let's pretend that you and I are all alone . . ."

They had already set a new endurance record, orbiting the earth over one hundred twenty times. Each new day brought them sixteen new sunrises and sunsets. Their first sunrise had filled both astronauts with silent awe. As the sun climbed through the earth's atmosphere, its light was split into vivid reds and blues and yellows. To Borman it appeared as if he was "looking into a huge cave with a red mouth, yellow roof, and blue outer rim."[16]

After more than a week, however, the sunrises and sunsets, while still beautiful, had become commonplace. Both men had lost weight, felt dirty, and were tired of the food. And now the much-anticipated rendezvous was delayed for at least three more days.

As a reward for their new endurance record, the ground finally relented and allowed both men to strip to their longjohns. "Hallelujah!" Lovell responded, immediately shedding his suit.

In Houston, Susan Borman and Marilyn Lovell were in many ways as uncomfortable as their husbands. Susan had taken her kids to Cape Kennedy to watch the launch, and had found the experience very frightening. Unlike military jets, which Frank could pilot and control, the Titan rocket looked more like a missile with her husband instead of a bomb in the nose cone. Worse, she and the children had always been insulated from Frank's test flights. When he flew a experimental jet she couldn't actually see him do it. He went to work and came home when the job was done. Here the violence and danger of his work was almost shoved in her face.

Gemini 7 lifted upward, its engines spitting fire and smoke and the roar engulfing the Borman family. Fred Borman, then only fourteen, gripped his mother's hand and asked, "Mom, why didn't you tell us it would be so difficult?"[17] Susan Borman held him tighter, not only to comfort him but to keep her own fears under control, knowing that all around them

Susan Borman, with Ed on left and Fred on right, watch launch of Gemini 7.

news cameras were clicking away. To the public she had to remain the supportive, excited wife.

Marilyn Lovell had also wanted to go to the Cape to watch her husband's first launch, but unexpected circumstances intervened. She had become pregnant. The Lovells already had three children, and hadn't planned on any more. Suddenly, in the late spring, with Jim already assigned to Gemini 7, Marilyn realized that she was going to have another baby, and would give birth either during Gemini 7 or immediately after. She decided to tell no one about it. She worried that NASA might take Jim off the flight because of her.

Of course, keeping this secret wasn't possible for long. When Jim and NASA found out, however, they did nothing. Jim stayed on the mission, and Marilyn prayed that news photographers would only photograph her from the neck up — which they did. Nonetheless, being nine months pregnant, the idea of traveling to Florida to see that rocket blast off seemed a very bad idea. Marilyn watched everything from her home.

Also watching from Houston was Bill Anders. After trying unsuccessfully for three years to become a test pilot, he had found a way to become an astronaut instead. One Friday afternoon in 1962 he was driving his Volkswagen bus home from work when he heard on the radio that NASA was now accepting applications for its third class of astronauts. Anders listened with casual interest. Up until now, all astronauts had been test pilots, and since he'd never even been to test pilot's school, becoming an astronaut seemed hardly a possibility.

The announcer began checking off the requirements. "Two thousand hours total jet time."

*I got that*, Anders thought.

"A masters degree in engineering."

*I got that*, Anders thought.

"A military career."

*I got that*, Anders thought.

Anders waited for him to mention test pilot training, but the announcer instead went on to describe how one obtained an application. To Anders's delight he suddenly realized that NASA no longer required its astronauts to be test pilots. Instantly he pulled to the side of the road so that he could write down the address for getting his application. When he got home he looked at Valerie and in his soft-spoken manner said, "Boy, wouldn't it be great to go to the moon?" He then told her he was going to become an astronaut.

Valerie's first reaction was "What?" She didn't oppose it, she just hadn't even considered the idea. She, like Bill, had assumed that test pilots' school had to be his next career step.

Her second reaction was "It'll be better than Vietnam." Already they knew men who had fought and died in Southeast Asia. Flying in space was a safer occupation. The goals of the space program also seemed far more worthwhile for both the country and for Bill, compared to the quagmire that even in 1962 Vietnam appeared to be.

Her third reaction was "Wow, that's really exciting!" Though she had always know that her husband had an adventurous spirit, flying to the moon was much farther than she had ever expected him to go.

Bill applied, and on his thirtieth birthday, October 17th, 1963, was accepted to NASA's third class of astronauts, fourteen members strong. The

Bill Anders and Mike Collins on jungle survival training in the Panama
Canal zone. Anders's clothes have been improvised from his parachute.

class included Buzz Aldrin, Mike Collins, Dave Scott, Dick Gordon, Gene
Cernan, and Al Bean, all of whom flew to the moon in later years.

The Anderses moved to Houston and built a home in El Lago, the
same development where the Bormans lived. In the two years since the
Bormans and Lovells had arrived in Clear Lake, it had become a thriving
community. The only evidence that Valerie saw of those once-empty farm
fields was the immediate lack of a supermarket. That too soon changed.

For a church, the Anders joined the small Catholic chapel at Ellington
Air Force Base. They liked Father Vermillion, the church was closeby, and
several of their neighbors, including astronaut Gene Cernan, belonged.

For the next five years Bill Anders worked as hard as he ever had, doing
whatever NASA asked him to do. He went on field trips to the deserts of
Nevada and the jungles of Panama (where he and Mike Collins captured and
cooked iguana for food). He made speeches to schools, community groups,

and colleges (where once he buzzed the famous "Chicken Ranch" brothel of Texas from his helicopter on his way to a commencement). He studied both the Gemini and developing Apollo spacecraft (taking thousands more hours of academic study). And he flew endless simulations.

But he didn't fly into space. His lack of test pilot experience, while not mandatory for NASA, had put him in the lower echelons of the astronaut corps, and it was four years before he was finally assigned to a flight, as back-up for Gemini 11.

During the Gemini years the closest he ever got to the real action was during the Gemini 8 mission in March 1966. Bill was one of the capcoms, and on his watch one of the capsule's thrusters began firing uncontrollably, forcing Neil Armstrong and Dave Scott to make an emergency return to earth. For a few short minutes Anders was at the center of the action, relaying information between the ground and the capsule as the astronauts regained control of their spacecraft and planned their sudden reentry. Then it was over.

Anders didn't give up. He pushed harder. Eventually he knew he'd fly: it was only a matter of time.

* * *

A preliminary inspection of Gemini 6's first stage rockets indicated that the engines had cut off because a small electrical plug had been jarred from its socket. Closer inspection revealed a second, more significant error. A dust cover, which in July had been placed on an engine valve during maintenance work, had mistakenly never been removed. This also would have caused Gemini 6's engines to stall, and would have done so in October — had they tried to launch then.[18]

At 7:37 AM (C.S.T.) on December 15th, after three tries, Gemini 6 finally took off. Six hours later Wally Schirra eased the capsule within ten feet of Gemini 7. The two ships now flew in formation, nose to nose.

Schirra looked across at Borman and Lovell, who after twelve days in orbit looked pretty disheveled. "Bluebeard, you don't have much of a mustache," he kidded Borman.

Gemini 6 approaches Gemini 7. Note the loose wires.

"Don't let them kid you," Borman answered in defense. "I'm just a blond."

Later Schirra pointed out some loose wires that were trailing from Gemini 7's rear. "You guys are really a shaggy-looking group with all those wires hanging out."

Borman's serious response was typical. "Where are they hanging from?" he asked instantly, worried about the integrity of his spacecraft. Schirra immediately got serious and carefully described the wires to him.

For four hours the capsules circled the globe together, taking pictures and joking with each other. "There seems to be a lot of traffic up here," Schirra noted when the many voices on the radio grew especially confusing.

"Call a policeman," the capcom answered.

At another moment Schirra, a Navy man, held up a sign saying "Beat Army" so that West Point graduate Borman could see it. Borman quickly flashed his own "Beat Navy" sign in response.[19]

Near the end of the rendezvous period, Stafford startled Borman, Lovell, and mission control when he suddenly announced "We have an object

The two spacecraft mere feet apart. Gemini 7 is slowly tumbling as
Gemini 6 approaches.

in view. Looks like it's in a polar orbit and in a very low trajectory, traveling
north to south."

The flight controllers in Houston jerked awake in alarm. Nothing should
have been coming at the astronauts from that direction. Were the Soviets firing
missiles at the two Gemini capsules? Was a meteorite racing towards them?

Stafford continued, "It looks like he's trying to signal us. Stand by —
we'll try to pick this up." There was a long, pregnant pause.

And then Wally Schirra began playing "Jingle Bells" on a harmonica,
accompanied by Tom Stafford with a string of bells.

After a few seconds the ground controller laughed. "You're too much,"
he told Schirra.

When asked how the two astronauts had smuggled this "unneeded"
equipment on board, NASA officials decided they really didn't need to know.
"I'm sure it wasn't a case of smuggling," one official rationalized.[20]

The two spacecraft broke formation, and after a little more than a day
in orbit Gemini 6 returned to earth, hitting the ocean only twelve miles from
the aircraft carrier *Wasp*.

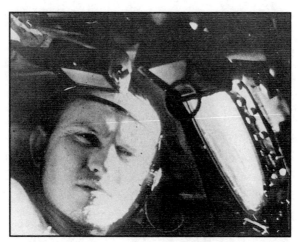

Borman near the end of the Gemini 7 mission. The hatch window is on the right.

Borman and Lovell, however, continued their confinement in space. After twelve days in orbit their spacecraft's operation was beginning to sag. The Gemini capsule (as would the Apollo spacecraft) used fuel cells to generate electricity. By now, however, two of Gemini 7's three fuel cells had failed. In addition, the two thrusters for controlling the ship's yaw no longer worked properly, and the ship's attitude control fuel was almost gone.

And the astronauts were tired, very tired. Borman, worried about the fuel cells, badly wanted to come home. At one point Chris Kraft, flight director, got on the radio to go over the problems and ease Borman's mind. When mission control noted that they only had three and half hours to go, Lovell responded, "Right-o. That carrier will feel good."[21]

Finally, at 8:28 AM (C.S.T.) on December 18th, the retro-rockets fired automatically, and after two hundred six orbits and more than five million miles, Gemini 7 came home.* And though for scientific reasons the doctors asked the two astronauts to stay in their capsule until it was hauled unto the

---

* Despite Gemini 7's immense total travel distance, it never rose higher than 203 miles elevation and was never far from home. Apollo 8, which flew one-tenth the total distance, traveled almost 1,200 times farther from the earth.

*Wasp*, both men refused. They insisted on being airlifted immediately by helicopter back to the aircraft carrier.

On the ground, both families were relieved. "I thought two weeks was an eternity, but the last thirty minutes seemed even longer," said Marilyn Lovell, describing how she had felt as the capsule fell to earth.

Then she returned to her role of supportive wife. "It was the most perfect mission I could have hoped my husband could possibly be connected with. He could come home, beard and all, and I would welcome him with open arms."[22]

Susan Borman also played her part. She looked at the televised picture of her husband on the deck of the *Wasp* and told reporters, "He looks marvelous. I think the flight was wonderful and great."[23]

Even as Borman and Lovell arrived on the Wasp, mission control in Houston was filled with cheers of celebration and triumph. Robert Gilruth, director of the Manned Spaceflight Center, noted that "it has been a fabulous year for manned space flight."[24]

In the next year, the United States launched five more Gemini missions, one every two months. Each was more successful then the last, achieving every goal and proving that humans could not only survive in space, they could work there as well. Furthermore, the next generation of American rockets, the Saturn 5, was rolling off the assembly line. This was the rocket that, if all went as planned, would take three Americans to the moon sometime in 1967.

During this same period the Soviet space program under Brezhnev was working non-stop to develop its new Soyuz space capsule. With this spacecraft they also hoped to fly two cosmonauts around the moon and back to earth by 1967.

If all went well, 1967 would finally be the year that both nations flew human beings to the moon.

GREGORY

Fred Gregory stared at his knees. Until his co-pilot slapped him on the shoulder to point them out, he hadn't realized they were shaking uncontrollably.

Gregory laughed. While his nervous system might have a mind of its own, *he* was having the time of his life.

It was 1966 and America was embroiled in the war in Vietnam. Fred Gregory was a helicopter pilot, and he now hovered about seventy-five feet above the jungles of South Vietnam. Below him burned the wreckage of a small reconnaissance plane, its pilot dead and its one passenger, a local scout, waiting desperately for rescue. Bullets were flying everywhere, and all around him American planes strafed the ground with cover fire.

Two years earlier, North Vietnamese gunboats had attacked U.S. destroyers patrolling international waters in the Gulf of Tonkin, off Vietnam.* President Johnson, having taken over as President after John Kennedy's assassination in November 1963, immediately responded with the first American bombing strikes on North Vietnam, proclaiming that "aggression by terror against the peaceful villages of South Vietnam has now been joined by open aggression on the high seas against the United States of America. The determination of all Americans to carry out our full commitment to the people and to the government of South Vietnam will be redoubled by this outrage."[25]

Shortly thereafter, Congress overwhelmingly passed what became known as the Tonkin Gulf Resolution. This law authorized the President "to take all necessary measures to repel any armed attack against the forces of the United States and to prevent any further aggression." It also gave Johnson the power "to take all necessary steps, including the use of armed force, to assist any member or protocol state of the Southeast Asia Collective Defense Treaty requesting assistance in defense of its freedom."[26]

The complexities and failures of the Vietnam War can hardly be analyzed here. What can be said is that few Americans at the time questioned the need for this military action. Like Berlin, it merely seemed another front in the war with communism, tyranny, and Soviet power.

Unlike Berlin, however, the war that President Johnson and Congress had so quickly decided to join was much more tangled. While Vietnam was partly an internal civil war between capitalist and communist factions, it was

---

* In later years it was learned that, while one gunboat attack did occur, a so-called second more serious attack almost certainly did not happen, even though this second attack was used by the Johnson Administration to justify the bombings and Tonkin Gulf resolution. Herring, 133-137; Karnow, 365-373; Moss, 156-165.

also a war of independence from colonial rule. And unlike Berlin and Europe, it appears now in retrospect that the faction that wanted a communist Vietnam was probably in the majority.

In the twenty-four months following passage of the Tonkin Gulf Resolution the U.S. contingent in Vietnam increased from 23,000 men to 170,000. The fighting had escalated, and the casualties had mounted, with little sign of progress or settlement.

During the build-up, Fred Gregory was an Air Force helicopter rescue pilot stationed at Vance Air Force Base in Oklahoma. Though his job in Oklahoma was to perform rescues, little ever happened. And because Gregory had to always be on call in case of emergency, he could never fly his helicopter more than three minutes from base. His situation reminded him of Henry Fonda's in the movie "Mr. Roberts." Just like the character in the movie, Fred could see the action in Vietnam passing him by, and this frustrated him.

Gregory was born to a middle class black family in Washington, D.C. which for generations had fought, and beaten, the bigotry of their time. His great-grandfather had been a member of Howard University's first graduating class. His grandfather had been a successful carpet layer and union organizer.

His father had graduated from both Case Institute (now Case Western Reserve University) and M.I.T. as an electrical engineer, but could not get work in his profession because of his skin color. Instead, he became a teacher, rising quickly through the Washington, D.C. school system to become its assistant school superintendent for vocational education.

When Fred was thirteen and about to enter the senior year of junior high school in Washington, the Supreme Court made its ruling against segregation in *Brown vs. Board of Education*. Though the nearest junior high school was only a short walk from his home, Fred had been attending a segregated black school halfway across the District.

The Washington school administration decided that they would desegregate all classes except for the seniors in junior high and high school. They assumed that it made less sense for kids about to graduate to change schools.

Fred's father thought it absurd for his son to have to travel so far each day, especially with a school so close to his home. Many of Fred's friends were white children who lived in the same neighborhood and attended the nearby school. Francis Gregory made the proper arrangements, and when school opened in the fall of 1954, Fred Gregory became the only black student in the ninth grade of John Philip Sousa Jr. High School.

For the first few hours, Fred was also the *only* student in the ninth grade of John Philip Sousa Jr. High School. He and his home room teacher sat together in class and stared out the window at the rest of the school's students, gathered on the grass of adjacent Fort Dupont Park to protest segregation's end.

It was a silly protest. Fred was simply doing what all the kids there did — attend his local neighborhood school. Since he was friends with many of those young demonstrators, the protest carried little steam. By lunchtime the kids were back in class, and without much additional fanfare the school quickly accepted integration.

To Fred and his family, such events were merely small hurdles on the way to success. They had always taken the attitude that the only obstacles you faced in life were obstacles you put there yourself. They had faith in American concepts of freedom and peaceful dissent, and despite facing the worst forms of racism, had seen their hard work pay off through four generations.

In many ways, Fred Gregory was very similar to the Apollo astronauts. Like Borman and Lovell, Fred Gregory had been interested in airplanes and flying since childhood. Like Lovell, Gregory had to attend a military academy (in his case the Air Force Academy in Colorado) in order to learn how to fly. And like Borman and Anders, he had a strong desire to serve his country.

Unlike the astronauts, however, the space program did not appeal to him. "I really didn't have a whole lot of interest in getting in a blunt body capsule," Gregory remembers. He wanted to fly planes, and "the thing didn't look like an airplane."

In 1965, however, he was stuck in Oklahoma, hardly doing any flying and watching the war in Vietnam unfold without him. He badly wanted to go overseas, and after making repeated requests for transfer, Fred finally got his wish in the spring of 1966.

Six months later he was holding that helicopter steady while his knees shook like crazy. In the back of the helicopter the two parachute jumpers (or

P.J.s) had lowered what they called a "forest penetrator" down to the stranded scout.

The penetrator was an elaborate harness which a soldier unfolded, assembled, and climbed into so that he could be safely hauled into the helicopter. Carefully-worded English instructions were attached so that anyone could figure out how to use it.

Unfortunately, the scout was not an American, and couldn't read English. Unaware that he was supposed to unfold the device, he just grabbed it and held on, expecting the rescue crew to haul him up.

This wouldn't work. They signaled for the scout to let go of the penetrator and pulled it back into the helicopter. One of the P.J.s opened it up and climbed on. The second slowly lowered his partner through the dense undergrowth and, acting as Fred's eyes, commanded, "A foot to the right . . . a little more . . . hold it . . . hold it . . ." as he yelled over the roar of the rotor blades and gunfire. The Vietcong were closing in.

The cable reached the ground. The scout, with directions from the P.J., climbed onto the harness seat, and both he and the P.J. hung on as they were raised back up to the safety of the helicopter. As the soldiers were hauled to safety, Fred's co-pilot tapped him on the shoulder and pointed at Fred's twitching knees.

They both started laughing. "It was really funny," Fred remembers. "I had absolutely no control over them." Though he was holding the helicopter rock steady, his knees were jumping about like popcorn in a popper.

Without waiting a second, and with his knees still shaking, Fred calmly pulled the chopper up and out of fire, heading north to the safety of the American airbase at Danang.


BERLIN

On December 19th, 1965, one day after Frank Borman and Jim Lovell returned to earth, West Berliners began lining up at the Berlin Wall. For the third year in a row, the East German government had agreed to allow limited visitation rights to any West Berliner with relatives in East Berlin. Beginning on this day and continuing for the next two weeks, West Berliners were allowed to make one or two visits to the Soviet Zone. Over 350,000 permits had been approved, and almost a million passes issued.

Hours before dawn and the opening of the gates, hundreds had arrived with their passes, carrying suitcases, shopping bags and boxes filled with gifts for their relatives in the East. One man brought a six-foot aluminum bathtub to give to his parents. On the East Berlin side anxious crowds formed as well. In all, over 60,000 West Berliners visited East Berlin on this first day, with 70,000 going on the next.

Dean Heinrich Grüber and his wife and daughter found themselves barred by the guards, however. Though the Grübers had passes, and merely wished to visit their son in East Germany, the guards denied them entry. Grüber was a prominent Evangelical church leader, and it was decided that his presence in East Berlin "was undesirable in view of the present political situation."[27]

On that same day, as the tens of thousands of West Berliners lined up to enter East Berlin, three East Germans arranged their own pass for *leaving* the Soviet zone. Since Peter Fechter had died trying to leap the wall in 1962, the methods of escape had become more creative. The East Germans had fortified the barrier significantly in the ensuing three years, adding a second inner wall, as well as trenches, watchtowers, and dog runs. To escape, refugees now built elaborate tunnels, some hundreds of feet long with lighting and tracks. Others designed secret compartments in their cars to conceal refugees. One East German stood on top of a building and threw a zipline across the wall and down to some West Berliners. Then he and his wife and nine-year-old son put on harnesses, hooked themselves to the cable, and slid down to freedom.[28]

In the four years since the wall's construction, a thriving cottage industry of professional escape-organizers had developed. Some did it for idealistic reasons, accepting just enough money to pay their costs. Others turned this work into an exciting but lucrative livelihood, earning significant sums of money.[29]

Horst Schramm was one of the professionals. A West German seaman who spoke English with an American accent, he had made about $400,000 smuggling refugees out of East Berlin. In the strange diplomatic universe of Cold War Germany, West Germans were permitted to enter East Berlin, but West Berliners were forbidden access. As a West German, Schramm took advantage of these rules to enter the Soviet Zone and arrange a variety of escapes.

On December 19th he drove into East Germany in a German-made Ford. This car closely resembled the vehicles used by the American Army in Berlin.

Once in East Berlin he picked up his three customers. Two were a couple in their twenties who had paid him about $800 for planning their escape. The third was the wife of a doctor, who had paid him an additional $1,000.

The doctor's wife climbed into the trunk of the car. The other woman was hidden in a secret compartment built into the car's dashboard. The two men, however, did not hide. Instead, they put on American Army uniforms that Schramm had "purchased" illegally. Schramm then replaced the car's license plates with a set of stolen American Army registration plates, and as a final touch, attached a tiny American flag to the front hood ornament.

The plan was simple. According to the original Four Power agreement signed by the U.S., Great Britain, France, and the Soviet Union at the end of World War II, officials from any of the four occupying nations had the freedom to travel anywhere within Berlin, and could not be questioned by any German police officer as they crossed Checkpoint Charlie. Disguised as Americans, Schramm expected that they would nonchalantly drive right through, unquestioned and more importantly, unsearched.

They reached the checkpoint. There the barricades were arranged to force a car to zigzag back and forth at less than ten miles an hour. As Schramm eased the car through, both men held up their forged identity cards and smiled at the East German guards. The guards in turn waved them through, having too much else to do that day. Thousands upon thousands of West Berliners were lined up for blocks, all waiting to get into East Berlin for a few hours.

The escape was easy. Unfortunately, the repercussions were not. When the story hit the newspapers, the U.S. Army, while "expressing sympathy" with the desire of the refugees to flee East Germany, condemned the unauthorized use of its uniforms and license plates. Within a week Schramm and the two U.S. soldiers who had sold him the uniforms were arrested.[30] The two G.I.'s were court-martialed and sentenced to three and four months of hard labor.[31] Schramm was fined $250 and sentenced to a six week suspended sentence.[32]

Three years earlier, Schramm and his cohorts would probably have been seen as heroes. Now, the response in the U.S. to their arrest and sentencing was a collective yawn, with this and other Berlin Wall escape stories hardly noted in the American press. On Christmas Day a man was shot to death when he tried to crash his car through the wall. That same week, two American soldiers were sentenced to eight years' hard labor by the East German government for their failed attempt to help an East German girl escape in September.[33] Neither story received more than passing mention in the West.

Things had changed since Peter Fechter's death in 1962. Then, even if the U.S. could do little to help, the national will had been strong and undiluted. "We stand for freedom," Kennedy had proclaimed, and Congress had unequivocally backed that proclamation by agreeing to a bold space race. If the U.S. couldn't tear down the Berlin Wall, or protect individual freedom in the Soviet Union, it could at least keep the Soviets from controlling outer space.

By 1965, that national will had changed, moving to other, more difficult conflicts. The problem of Berlin was no longer front page news. The space race had begun to lose its sheen and glamour. Fred Gregory, who loved flying and was so similar in mind and spirit to the astronauts, cared very little about their space program. Instead, he wanted to get to Vietnam.

The day that Frank Borman and Jim Lovell began their two week orbital mission, the *New York Times* reported that the Johnson Administration was planning in the next year to increase its troop strength in Vietnam from 170,000 to 400,000 men. In the last two months alone, U.S. and South Vietnamese forces had suffered more than 8,500 casualties.[34]

Only six days earlier approximately 25,000 people had gathered in Washington to protest that war. For two hours they demonstrated in front of the White House, then marched to the Washington Monument where they listened to speeches from, among others, former Socialist Party Presidential candidate Norman Thomas and baby care specialist Dr. Benjamin Spock. "We should turn Vietnam over to the Vietnamese people for them to decide their government as they see fit," said Spock.[35]

Nor was this the only issue that people were protesting. Earlier in the year there were riots in black neighborhoods in Los Angeles and Chicago. The Los Angeles riots, in the neighborhood of Watts, lasted almost four days, claimed over thirty lives, and required more than 20,000 National Guardsman to quell.[36]

Even as Borman and Lovell circled the world several hundred times, the world had been turning under them. It was now turning in directions no one had predicted or understood.

CHAPTER SEVEN

# "HEY, I GOT
# THE MOON!"

ALONE. The three men were now more alone than any humans in history. They had spent the last three days watching the blue-white planet of their birth dwindle behind them. After the first two days of travel, Lovell wasn't sure if he was ebullient or anxious when he realized he could cover the entire planet with his thumb.

Now their spacecraft had passed behind the moon, cutting them off from contact with earth. In mere seconds their S.P.S. engine would ignite and place them in a stable, lunar orbit — or so they hoped.

* * *

Since ending their Monday press conference twelve-and-a-half hours earlier, the three astronauts had spent most of that afternoon resting and making last preparations for this moment.

Lovell did some additional navigational sightings, estimating numbers for several future mid-course corrections as well as the burn that would put them into lunar orbit. By now he was getting so good at using the computer and the sextant that he felt "like a concert pianist" as his fingers played across the computer keyboard.

Anders continued monitoring the spacecraft's health. At one point he decided to use the on-board tape recorder to report some miscellaneous facts to the ground, narrating a list of minor problems onto the tape. "The food box doors are hard to close . . . Look's like we've gotten the handle bent in trying to close the door . . . The meals I've had have been quite tasty, though none of us have really gone overboard for the little bread cubes and cereal cubes . . . If they ever fly one of these TV cameras again, they [should] put some sort of sight on it . . . Tell Doc Frome that his toothpaste tastes pretty good. I don't know what kind of job it does on your teeth, but it's nice for settling your stomach after dinner."

Lovell then quipped in the background, "We used it for frosting on the fruitcake."

Anders continued his report, "Jim Lovell is . . . engaged in an activity which I shan't describe, so I think I'll cut this short and get my oxygen mask."

"But that could be improved also," Lovell added, referring to what NASA euphemistically called the "Waste Management System." The tape then ended with all three astronauts laughing like adolescent boys.

At about 7 PM they did their last mid-course correction. This time, rather than the S.P.S., they used the four clusters of attitude control jets on the sides of the spacecraft's service module. These smaller engines, each with a thrust of one hundred pounds (compared to the S.P.S.'s 20,500 pound thrust) fired for eleven seconds, putting the spacecraft to within a ten thousandth of a percentage point of the planned course. The astronauts would now slip behind the moon at 3:50 AM Tuesday morning, swinging past it at the desired distance of seventy miles. If all went according to plan, at 3:59 AM they would then fire the big S.P.S. engine for four minutes. This burn, called Lunar Orbital Insertion (or L.O.I. for short), would slow the spacecraft's speed from 5,300 to 3,700 miles an hour and put it in lunar orbit. If for some reason the astronauts decided not to fire the engines, they would whip around the moon and be flung back to earth, where they could land in the ocean as had all other American space flights.

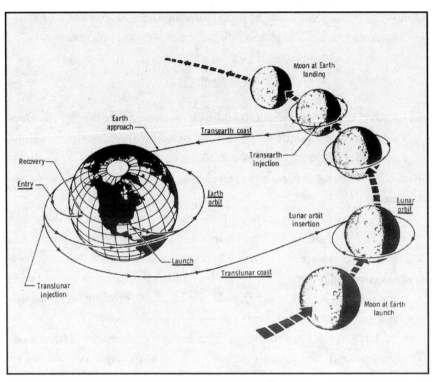

The Apollo 8 mission profile.

If the S.P.S. worked as planned, however, the astronauts would become the first human beings to not only escape the gravitation pull of the earth, but to join the environment of another world.

For the rest of that Monday evening the men mostly rested. Anders took a sleeping pill. From 10 PM to 1 AM almost nothing was said between the spacecraft and the ground. Long periods of silence were broken by short conversations. Twice Jerry Carr passed up some numbers, once to Lovell and once to Borman.

Yet none of the crewmen slept much. As Monday turned to Tuesday and Christmas Eve arrived on earth, these three space travelers were more than 220,000 miles away from home. With each second their speed was increasing, the moon pulling at them, drawing them in. As the clock struck midnight, their speed was over 3,000 miles per hour, and increasing steadily.

\* \* \*

As Marilyn Lovell awoke, she could hear the television faintly, its sound drifting in from the family room. Yet she heard no voices or the movement of people. Her bedroom clock said 2:00 AM in the morning. The spacecraft would be entering lunar orbit very soon, and she wondered what was happening.

Quietly she crept from her room, which was located on the lower level of her split-level home. She peeked into her childrens' rooms, finding that Jeffrey and Susan were fast asleep. Then she poked her head up the stairs to look into the family room. There the floor and chairs and sofas were covered with the sleeping bodies of her friends, the television droning on about how Apollo 8 was about to arrive at the moon.

Marilyn stared in wonder and humility at the generosity of her friends. They had put her two younger children to bed, then stayed on, waiting for her to wake up. She made a noise to let them know she was coming, and joined them in the family room.

\* \* \*

Valerie had slept, but unlike Marilyn she hadn't needed any urging. She was riding such a high about the space flight that the dangers seemed somehow unreal to her. "Maybe it was a form of self-hypnosis," she remembered years later.

Possibly more significantly, she had spent most of the evening getting her five very young children to bed, telling them stories and helping them say good night prayers. Each had many questions about what their father was doing, and she spent a good deal of time trying to explain it to them.

Then she went to sleep, and only awoke when a few friends arrived to be with her for L.O.I. She got up and together they gathered around the television to watch as her husband reached the moon.

\* \* \*

Finally, after almost three days and 240,000 miles of travel, Apollo 8 had arrived in lunar space. The men were now so far from the earth that radio communications, traveling at the speed of light (186,280 miles per second), took two and a half seconds to go from the capsule to Houston and then back to the spacecraft. The ship's speed was over 5,000 miles an hour, and the distance to the moon had shrunk to mere miles. Yet, Borman told the ground, "As a matter of interest, we have as yet to see the moon." Their spacecraft's orientation — tail pointing to the moon — still prevented them from seeing it.

The unseen world that these three men now approached had tantalized humanity across thousands of generations. Civilizations had come and go, each watching the moon wax and wane as its perpetual lunar cycle clocked the passing of the seasons, each culture trying to understand the origin and substance of this glowing silver-white sphere in the sky. The Navajos believed that the First Man and First Woman made the sun and moon to brighten the world with light, and used a slab of quartz crystal to carve the disks, attaching them to the sky with lightning darts. The Greeks, while believing in the moon goddess Selene, also insisted that the moon must be a planet like the earth. "The moon appears to be terrestrial," said Plutarch, "for she is inhabited like the earth . . . and peopled with the greatest living creatures and the fairest plants." Many cultures told of a Man-in-the-Moon, and because new or green cheese resembled the moon, some legends even jibed that maybe that was what constituted the moon.[1]

With the coming of the telescope, astronomers learned that the moon was truly another world, but airless and almost certainly without life. It had a diameter of 2,160 miles, making it about a quarter the size of earth. In the seventeenth century, astronomer Giovanni Riccioli studied the surface and named many of its most prominent craters and mountain ranges. He also named the large dark areas of the moon *mare* (pronounced MAR-ray), Latin for "sea" because he thought these regions were either oceans or dried seabeds.

Despite centuries of careful astronomical observations, however, no human being had ever seen the moon's far side. Because the moon's day (27.32

earth days long) is exactly the same duration as its orbit, the satellite always presents the same face to the earth. The far side is forever turned away, though orbital wobbles make about sixty percent of the entire surface visible over time. The remaining forty percent had remained eternally veiled, a tantalizing mystery just beyond reach.

In the nineteenth century science fiction writers made fanciful guesses about what might be hidden on that unseen hemisphere. In his book *From the Earth to the Moon,* Jules Verne had a scientist propose that the moon was shaped in "the form of an egg, which we look at from the smaller end." Because most of the moon's mass was therefore hidden from our sight, this fictional astronomer proposed that the heavier gravitational field on the far side allowed the moon to retain air and water. On the far side, life existed, and there a lunar explorer could survive.[2] And while most later writers, unlike Verne, assumed that the hidden lunar surface resembled the near side, all hoped somehow that concealed in the moon's secrets were alien species and ancient civilizations.

Finally, in 1959, the Soviets sent Luna 3 on a fly-by mission past the moon, obtaining the first glimpse of the unseen hemisphere. Though the pictures were poorly resolved, computer enhancement revealed a mountainous rough surface with only two dark areas resembling the near side's mare regions. Then, in the mid-1960s, the Soviets and Americans launched a total of eight lunar orbiters, mapping the far side more thoroughly and unveiling a surface riddled with craters.

None of these photographs, however, could compare with what Borman, Lovell, and Anders would see when they slipped into lunar orbit only seventy miles above its surface. Not only would the astronauts see terrain previously unseen by humans, they would see it in three dimensions, adding a reality imperceptible to any robot ship photograph.

At 3:00 AM Jerry Carr took a breath and announced calmly, "Apollo 8, this is Houston. At 68:04 [hours into the flight] you are *go* for L.O.I."

Borman said, "Roger. Apollo 8 is *go*."

Carr answered, "You are riding the best bird we can find."

Borman: "Thank you. It's a good one."

What Borman and the other two astronauts didn't know was that for the last three days ground engineers had been trying to figure out why the

S.P.S. engine had not performed exactly as anticipated on Saturday's mid-course correction. While the engine had worked, subsequent analysis showed that it had generated slightly less thrust than expected. After what the engineers later described as "intensive discussions," they concluded that the lost thrust was due to helium bubbles trapped in the fuel lines. After the mid-course correction burn, the engineers felt that the lines were now bled clear and that the S.P.S. engine would work as designed during L.O.I. On this recommendation, the men in mission control took a deep breath and cleared Apollo 8 for lunar orbit.

Five minutes before the astronauts slipped behind the moon, Jerry Carr relayed a message from Susan Borman to Frank. Carr cryptically said, "Frank. The custard is in the oven at 350."

Borman didn't get it. "No comprendo," he told Carr.

In the early days of their marriage, Frank and Susan had worked out what they each needed to do to make their partnership work. They both agreed that while Frank's job would be to work and pay the bills, Susan would run the house and raise the kids. Frank summed up this agreement by saying that he "would fly the jets and she would cook the custard."

He was not being flip or insulting. As hard as he tried, he couldn't have succeeded without Susan. He needed her to make the family work, not only for himself but for his children.

It was a job that Susan loved anyway. For years Frank's remark stuck in her mind, a neat metaphor for the symmetry of their lives. With Frank flying the most powerful flying machine ever built, Susan used it to let him know that she was also doing her part.

Frank, however, had forgotten the phrase. Not that it mattered. He knew, without anyone telling him, that Susan was there for him. On earth, she waited in her kitchen for her husband to disappear behind the moon. It didn't bother her that Frank hadn't understood. She preferred he stayed focused on what he needed to do.

At thirty seconds before loss of signal, Jerry Carr said, "Safe journey guys."

Anders answered, "Thanks a lot, troops."

Lovell added, "We'll see you on the other side."

Carr: "Apollo 8, ten seconds to go. You're *go* all the way."

Borman: "Roger."

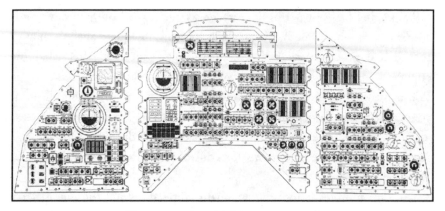

The command module's instrument panel.

And then silence. At ten minutes to four in the wee morning hours of Christmas Eve 1968, their ship passed behind the moon, cutting off all radio communication with mission control in Houston as well as every other person on earth.

* * *

On the spacecraft, both Borman and Anders were astonished at how precisely the computers had predicted the loss of signal. Borman said, "That was great, wasn't it? I wonder if they turned [the transmitter] off."

Anders laughed. "Chris [Kraft] probably said, 'No matter what happens, turn it off.'"

It was T minus ten minutes before L.O.I. Borman and Anders began going down their checklist, Anders reading off a particular setting and Borman checking the instrument panel and confirming aloud that the setting was correct. There were several hundred switches on the instrument panel, and they all had to be right. The astronauts oriented the spacecraft. They checked, and double-checked, to make sure the ship's manual controls were linked to the computer. They configured the spacecraft's circuitry. They checked the spacecraft's pitch, roll, and yaw.

Lovell, after programming the computer, had little to do now but wait it out. He glued his eyes to the windows and stared at the black sky. Still no moon. He could see the stars — the sky was littered with them. Because this flight was a scouting mission for the planned lunar landing several months hence, NASA had scheduled it so that the Sea of Tranquility, the prime landing site, was close to lunar sunrise, thereby accentuating the shadows and making it easier for astronauts to pick out details. This schedule, however, required Apollo 8 to plunge towards the moon on its night side, the spacecraft traveling through the moon's shadow. Moreover, when they slipped behind the moon the earth with its earthshine was also cut off. They were now in the darkest lunar night, surrounded by an infinity of stars.

At T minus eight minutes, Anders and Borman had finished the first part of their checklist. Now they had a few minutes to wait before beginning the final countdown to the S.P.S. burn.

The men sat in silence. Anders gazed out his window at the sea of stars, still not having seen the moon. Suddenly a chill ran down his spine. Across that star-flung vault of heaven now crept an arched blackness, a growing void within which he could see no stars at all. The moon was approaching.

After almost forty seconds of quiet, Lovell spoke up. "Well, the main thing to be is cool."

"Gosh, it is cool," Borman answered.

Lovell looked at the cabin thermometer. "It's up to eighty [degrees] in the cockpit."

Anders tried to explain how they felt. "No, I think . . . just when my clothes touch me, it gets cold, huh?"

At T minus six minutes, Borman said, "Okay, let's go," and he and Anders began going down their final checklist, setting the last switches and arming the engines.

At T minus 2:20 Borman glanced out the window. According to their calculations, the sun would be rising on the lunar horizon any second. "Boy, I can't see squat out there."

"You want us to turn off your lights to check it?" Anders suggested.

Lovell cut in. "Hey, I got the moon!" With a bright flash the sun rose, casting long streaks of harsh light across the lunar surface below them.

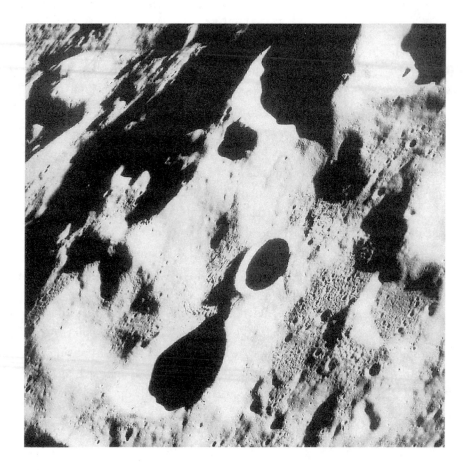

"Do you?" Anders asked.

"Right below us," Lovell said.

"Is it below us?" Anders leaned towards his window. Though he had already seen the long bands of light cutting across the lunar surface, his mind hadn't yet comprehended what these were. For a moment he thought — though he knew this was impossible — that they were streaks of oil running down the window.

"Yes, and it's — "

"Oh my God!" Anders gasped as he fathomed what he saw. He was staring at a black-and-white surface of mountains and craters, suddenly so near.

"What's wrong?" said Borman, frightened that the man in charge of monitoring spacecraft systems had discovered a serious mechanical failure.

"Look at that!" Anders said with wonder.

Borman looked outside, saw the moon, and struggled with a desire to stare like the rookie. Then he thought of the impending engine burn and pulled Anders back. "Stand by . . ."

Anders read off four more items on the checklist, with Borman confirming them. Then they all paused. T minus 1:50 seconds to L.O.I.

Anders stared out the window again. "I see two . . . Look at that . . . Fantastic!"

"Yes," said Lovell.

"See it?" Anders continued. "Fan — fantastic. But you know, I still have trouble telling the holes from the bumps."

Borman cut in. "All right, all right, come on." There was only a little more than a minute to go. "You're going to look at that for a long time."

"Twenty hours, is that it?" Anders said. He went back to his flight plan and began reading aloud the last commands. "Standing by for engine on enable." To Lovell he said, "Proceed when you get it." This was an instruction that Lovell needed to input into the computer.

Lovell: "Okay."

Anders: "Start your watch when you get ignition." There were now less than four seconds to go. "Stand by for —"

The S.P.S. engine fired automatically. Lovell called out, "Enabled!" and the three men were pressed back against their couches.

A quarter of a million miles away, an entire world waited in breathless suspense. In Houston Jerry Carr sat at his console. He called, "Apollo 8, Houston, over," waited fifteen seconds, and then called again, "Apollo 8, Houston, over." Again and again he did this, despite pronounced mixed feelings, knowing that if he regained communications before 4:30 AM, something would be terribly wrong.

In Moscow, the Soviet government newspaper, *Izvestia*, had described this moment by noting that "the slightest miscalculation might make the astronauts forever captives of the moon."[3] Having led the space race from day one, the Soviet Union was finally taking a back seat to an American achievement in space.

Susan Borman sat silently, alone at her kitchen table, hands clasped and head bowed, listening intently to her squawk box. Close by in the den her two teenage sons waited, watching the television with several friends, Frank's parents, and Reverend James Buckner from their parish.[4]

At Marilyn's, about a dozen close friends, her two oldest children, and Father Raish sat quietly in the family room. Periodically someone would try to start a conversation, but the words would fade out after a sentence or two.

Valerie Anders, however, was hardly awake. "I just didn't think L.O.I. was that dangerous," she remembered. She was also aware that in a few short hours her swarm of children would be getting up, and to be ready for the next day's challenges she needed her sleep. Almost nonchalantly and with mild impatience, she waited with her friends for word that everything was all right so that she could go back to bed.

Right now, however, the only thing anyone could hear was Jerry Carr's patient voice.

For four long minutes Apollo 8's engines roared. All three men knew that if the rocket cut off too soon, they would be put into an erratic lunar orbit from which they might not have enough fuel to escape. And if the rocket burned too long, they would instead be killed as their craft crashed on the moon. Such a landing was not how the United States wished to win the space race.

More likely and much more worrisome, however, was the possibility that their position was not as they had calculated it. They had traveled 240,000 miles, and should their path through space have been off the slightest fraction of a degree, they could be passing the moon at a distance either much closer or much farther than predicted. The rocket might then fire exactly as they had programmed it, and merely fling them against the lunar surface.

Unfortunately, they wouldn't know if this could happen — until it did.

After two full minutes Lovell asked, "Jesus, four minutes?"

Borman shook his head. "Two minutes." He looked at Anders. "How's it doing, Bill?"

"Great shape. Pressures are holding. Helium's coming down nicely. All other systems are go."

Because there was no atmosphere surrounding the spacecraft, the S.P.S. rocket made little sound. The astronauts could hear a hum, and feel a vibration through the spacecraft's hull, but other than that there was silence.

Note the black lunar horizon on the upper right.

After three minutes Lovell said, "Longest four minutes of my life."

Finally, Anders began counting down the last few seconds. Though their computer was programmed to shut the rocket down on schedule, Borman pressed the cut-off switch, just to make sure. Once again the astronauts were weightless, and the craft was silent except for the sound of their breathing. A quick check of the computer showed that they were now in lunar orbit exactly as planned, an eliptical orbit 69 by 194 miles high. "Congratulations, gentleman," said Anders. "You're at double zero," meaning they had hit their target precisely.

Borman thought how far from home they were. "Well, now is no time for congratulations yet."

Lovell grinned. "No, we get stuck with that on the carrier."

Still, they were alone, cut off from earth for another twenty-five minutes. Anders and Lovell stared out the windows. Below them drifted a desolate, crater-pocked surface, absolutely without color. Craters piled on craters piled on craters. Worn mountain peaks rounded by eons of impacts. Anders laughed. "It looks like a big beach down there."

Borman, still focused on getting them home, asked for the flight plan.

Lovell pulled it out, opened the book, and joked, "Holy cow, it's completely blank here."

The men went back to work, scrambling to set up the cameras and snap as many pictures as possible. For twenty minutes little was said as they raced to photograph that stark moonscape. Should they be forced to leave lunar orbit on the next pass, these photographs would guarantee their flight had some results.

* * *

It was now 4:29 AM (C.S.T.) on earth. The men in Houston's mission control waited, barely able to speak or breathe. In the homes of the astronauts everyone watched the television with apprehension. The only sound was Jerry Carr's voice, repeating his prayer-like litany into the radio, calling for Apollo 8.

Then, just as predicted, the control room reacquired signal from the spaceship, confirming the successful orbit insertion. Someone excitedly yelled, "We got it! We got it!" and then Jim Lovell's voice responded, "Go ahead, Houston." Amid the cheers and shouting that now filled mission control, Lovell dryly gave them a report on the engine burn. "Burn on time, burn time four minutes, six and a half seconds."

In the Anders home Valerie listened, heard they had gotten into lunar orbit, and then went back to bed. It was all very exciting, but her young family would need her well-rested when they woke in the morning.

In the Lovell home there were cheers and screams of joy. Marilyn was honored that Jim's voice was the first ever heard from lunar orbit.

At the Bormans there were also cheers, mostly from the boys and from Frank's parents. Susan closed her eyes and sighed. Now they were in lunar orbit. She tried not to think about what she *knew* would happen next.

\* \* \*

For the next hour and twenty minutes, until they disappeared again behind the moon, Jim Lovell and Bill Anders excitedly reported their view to a breathless earth. While Anders moved from window to window, snapping pictures, Lovell did most of the talking, trying to describe what he saw. "The moon is essentially gray, no color," he said. "Looks like plaster of paris or sort of a grayish beach sand."[5]

Borman, meticulous test pilot that he was, said little. He worried more about the state of his spacecraft than what the moon looked like at seventy miles elevation. After keeping quiet for about a half hour, he finally spoke up. "While these other guys are looking at the moon, I want to make sure we have a good S.P.S. How about giving me that report when you can?"

This brought laughter in the control room. Everyone knew how seriously Borman took his job, and somehow, in the excitement of the moment, it seemed funny that they could still depend on him to get it right. Jerry Carr answered, "Sure will, Frank."

Borman added, "We want a *go* for every rev[olution], please. Otherwise we'll burn in T.E.I. one at your direction." T.E.I. stood for Trans-Earth Injection, the rocket burn that would take them out of lunar orbit and back to earth. Borman's statement meant that if he didn't get an okay to remain in lunar orbit, he would assume they wanted him to come home immediately.

Borman was gently but firmly telling everyone on the ground to get back to work. He wanted to focus everyone's minds on the needs of the mission. If the planned rocket burn should fail twenty hours hence, they would live out the few remaining hours of their lives as prisoners of the moon.

And, if it was within his power, Borman did not intend to let that happen. He knew how close death lurked.

## CHAPTER EIGHT

# "SETTLE THIS BY NIGHTFALL."

## APOLLO 1

THE APOLLO CAPSULE WAS A CHARRED AND ASHEN RUIN. The smoke was gone, and the air was clear, but the command module held unmistakable signs of violent and sudden death. Parts of the control panel, blackened with soot, had actually buckled from the heat. Bits of burned insulation peeled from the walls and melted wires dangled everywhere. To one side hung the scorched inner hatch, its edge still fused to the capsule's wall.

Frank Borman, along with the nine other members of the launchpad fire review board, peered into the space capsule and saw instead a horrible oven. Twenty-four hours earlier, on January 27th, 1967, three fellow astronauts had died here.

Gus Grissom, Edward White, and Roger Chaffee had been participating in a simulated launch countdown. Their planned flight, the first launch of NASA's new Apollo spacecraft, was only four weeks away. The men had put on their space suits and rode the gantry elevator to the top of the Saturn rocket. There they climbed into their Apollo command module and were sealed inside, just as if this were an actual launch. For five hours they rehearsed the countdown sequence with the launch team, dealing with the small problems expected from a new spacecraft.

At 6:31 PM, the count was stalled at T minus ten minutes while both astronauts and engineers wrestled with a continuing communication problem. Suddenly one of the astronauts commented that he smelled fire, and then Roger Chaffee cried out, "We've got a fire in the cockpit!"[1]

Within thirty seconds the flames engulfed the spacecraft, rupturing its hull and sending acrid smoke billowing above the launchpad. The heat was so intense that it fused the astronauts' suits to the capsule's plastic interior. It would take almost two hours to cut the men's bodies free and remove them.

Frank and Susan Borman and the boys had left Houston that weekend, going to the country lake house of Jim and Margaret Elkins to celebrate the tenth birthday of the Elkinses' daughter. It was a place where the Bormans could have some time off, where no one could find them.

Yet, NASA did find them. A sheriff knocked on the door and told Frank that he was needed immediately. Borman called the Manned Spacecraft Center and Deke Slayton, head of the astronaut office at NASA, gave him the bad news. Within minutes he and Susan were driving back to El Lago.

As they raced south in the dark Texas night, the open fields around them and the high sky above them, they talked about the accident. Frank couldn't help thinking aloud, "If Ed [White] couldn't get that hatch off, no one could." Suddenly a terrible seed of doubt was planted in Susan's mind: Frank, as indestructible and sharp as he was, could just as easily have died in that accident. Then they arrived at the home of Pat White, where the astronaut's wife had already received the horrible news from Bill Anders.

Two and a half years before, astronaut Ted Freeman had been killed in a plane crash. To everyone's horror, the first person to tell his wife about her husband's death had been a news reporter, knocking on her door to get some quotes.

Since then NASA had been much more careful. If someone was killed, the nearest astronaut would be located and immediately sent to the wife's home to tell her the bad news. If the nearest astronaut was more than twenty minutes away, the nearest astronaut wife, usually a neighbor within

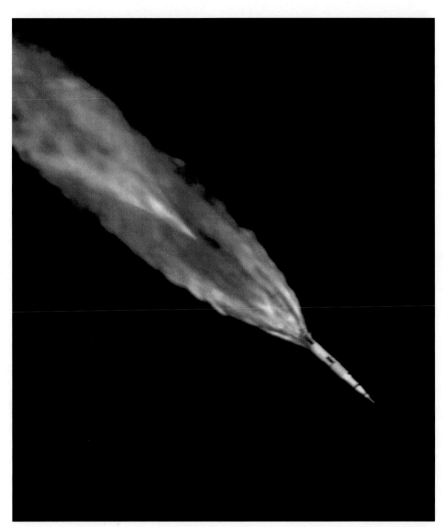

The Saturn 5 less than two minutes after launch, taken at 35,000 feet by a telescopic camera mounted on the cargo door of a C-135 aircraft. Because the rocket was already more than thirty miles high and over six miles down range, the camera looked almost directly *up* to take the photograph. Therefore, hold the book over your head when you view the picture and the angle of the Saturn 5 will then make sense.

Unless otherwise credited, all photographs and diagrams are courtesy of NASA.

The astronauts' first view of the earth after T.L.I. The Florida peninsula and Cape Canaveral can be seen in the upper left. Directly below this the Bahamas are visible, surrounded by light blue water. Further down the Caribbean island chain can be identified.

The earth progressively shrinks. The top left picture was taken mere minutes after the photograph on the previous page. Florida and the Bahamas can still be seen in the upper center of the globe, under the same cloud formations. Now, however, all of South America has come into view, with Chile and the continent's southerly regions pointing off the earth's left edge. Africa can be seen on the lower right. The top right picture was taken sometime late Saturday, and the bottom left picture sometime on Sunday. The bottom right picture was taken during Monday afternoon's television broadcast, and matches the televised view shown the earth on page 99.

Anders: "Looks like a sandpile my kids have been playing in for a long time. It's all beat up, no definition. Just a lot of bumps and holes." The only color, a very slight bluish hue, comes entirely from sunlight.

Earthrise. Compare to Frank Borman's photograph, page 173, taken a few minutes earlier.

December 27, 1968. After splashdown Frank Borman addresses the crew of the *U.S.S. Yorktown*, with Bill Anders and Jim Lovell standing beside him.

Looking into the hatch of Apollo 1 on January 28, 1967, one day after
the fire. All that remains of the center couch, where Ed White had sat, is
its metal frame.

walking distance, would be called as well to keep her company until the
astronaut arrived.[2]

Now Bill Anders, who happened to be home at the time, was the
astronaut assigned to give the news, and Jan Armstrong was the astronaut's
wife assigned to keep her company.

When the Bormans arrived shortly thereafter Susan was shocked by
what she saw. She and Pat had both seen their share of death: after all, their
husbands were military test pilots, and both had witnessed the death of many
other women's husbands.

Yet things were different that night. Three decades later Susan can still
remember how the feeling of disbelief and catastrophe that permeated that house
"unnerved me and unglued me like I had never been unnerved or unglued before."

That the men had died on the ground certainly made their death
harder to take. Test pilots died while in the air, which was what all the

astronauts and their families braced themselves for. No one at NASA had prepared them for an accident on the ground.

That one of the dead men was Ed White also made it more difficult. To many of the NASA community, including Susan, Ed was the astronaut's astronaut. He was strong, clearheaded, and able to solve any problem. Yet he had been killed in mere seconds, on the ground.

The grief was further amplified because some government official in Washington was already telling Pat White how to run the funeral. Ed had always told Pat that he wanted to be buried at West Point. Yet, try as she might, she was unable to make anyone understand that. The officials in Washington had immediately called to tell her that the government wanted all three men buried together, to make a "statement." They insisted that Ed would be buried at Arlington National Cemetery.

Frank stepped in. Almost as soon as he walked in the door he called the bureaucrats and told them in no uncertain terms that Ed White was going to be buried at West Point, just as he and his family had wanted.[3]

Then Frank was gone. Deke Slayton wanted him to represent the astronauts on the investigation committee NASA was forming, and before the evening was over he was piloting a plane to Florida. The board was going to inspect the Apollo 1 capsule that next morning.

In the weeks that followed, Susan and Pat spent a great deal of time together. The two women were very much alike, and had become friends almost as soon as they had met. They both loved their husbands and would have done anything for them. And they both were gladly and completely dependent on them for leadership.

Each evening Susan would go to Pat's home, trying to comfort her. But Pat was inconsolable. She had not anticipated Ed's death. He had been her pillar and her strength and, as Susan had with Frank, she had built her life around him. Now he was gone, and over and over again she asked Susan, "Who am I? What do I do now?"

Susan walked home across the suburban lawns and found herself asking the same questions. *Who am I?* she asked. *What would* I *do if Frank died?*

She had no answer. A test pilot's wife simply never thinks about the possibility that her husband could die. To think about the negatives only defeats you.

Borman testifying before Congress, April 17, 1967. Sitting next to him, from left to right, are astronauts Jim McDivitt, Deke Slayton, Wally Schirra, and Alan Shepard.

Yet, from this moment on Susan Borman began thinking about the risks Frank was taking. She began thinking that he could be next, and that she would be sitting where Pat White was now. Her faith in NASA and the space program was shattered. She resolved to keep this doubt secret, however. She told no one, including her husband.

For the next three months, as the review board investigated the fire, Frank practically lived in Florida. The three astronaut couches were carefully removed from the capsule and replaced with a wooden platform to allow panel members to get inside the capsule without disturbing the evidence. Borman sat in the command module, surrounded by what he subsequently described as a "fire-blackened charnel house,"[4] carefully studying every detail in order to discover the cause of the fire. Later, Borman repeatedly played the communication tapes of the accident, listening to the anguished scream of Roger Chaffee crying, "We're burning up!" in an effort to find out what happened.

The astronauts' deaths destroyed the lives and careers of several people at NASA. One man had a nervous breakdown. Borman, however, saw it no differently than the many other deaths he had witnessed while he was a test pilot. This was risky work, and people sometimes died doing it.

Nonetheless, Borman felt himself getting increasingly angry. Everywhere he and the rest of the investigation committees looked, they found sloppy workmanship by both the contractor and by NASA.

When Borman had been a freshman at West Point, something had happened that would define the rest of his life. The plebe's life was brutal: up at dawn, working sixteen hours a day, little time off, and little sympathy from anyone. One day the cadets had been running the plebes around all day, doing twenty-mile hikes, endless drills, and continuous workouts.

The day ended with bayonet drills on the West Point grounds. The plebes made repeated forward thrusts with their bayoneted rifle, holding it out at full stretch as if they had just stuck it in the gut of an enemy soldier. Never a big man, Borman only weighed one hundred forty pounds when he was eighteen. Now he was literally trembling with exhaustion. He couldn't hold the rifle steady.

Observing the plebe with contempt was a small, wiry lieutenant colonel who was actually a bit shorter than Borman. The officer came up to Frank and began screaming at him, telling him that he couldn't hack it, that he wasn't good enough.

Borman stopped trembling. He grew both calm and angry and looked that "little weasel" in the eye. No bastard was going to tell him he couldn't do something. He might be a naïve, country boy suddenly thrust into the hard, competitive world of West Point, but they'd have to kill him to get him out of the Academy. It was the same determination and strength of will that allowed him to bring every plane he ever flew safely back to earth.

Now, twenty years later, it was the same thing. Borman decided that he was going to do whatever it took to make sure the Apollo spacecraft flew again. And when it did, it would be the safest spacecraft ever built.

By mid-April, only ten weeks after the fire, the review board completed its investigation. More interested in solving the problem than laying blame, they made several recommendations, all of which were quickly adopted.

The Apollo capsule hatch would be redesigned. The old design had actually been two hatches, an inner hatch and an outer one, removable only by laboriously unscrewing some lug nuts. No one could have opened it in less than ninety seconds. The new hatch could be opened in under ten.

The use of a pure oxygen atmosphere before reaching orbit would be abandoned. At sea level the capsule pressure was set at 16.7 pounds per square inch, slightly higher than atmospheric pressure. This positive pressure isolated the capsule environment from the outside environment. Once in orbit the capsule pressure was then reduced to 5.5 pounds per square inch. At 16.7 pounds per inch pressure, however, pure oxygen is a lethal incendiary. By introducing a mixed atmosphere of forty percent nitrogen on the ground, the oxygen was diluted. As the capsule pressure was lowered to 5.5 pounds per inch in orbit, the nitrogen would be purged.

The use of combustible materials inside the capsule was to be severely reduced. More than 2,500 different items were removed, replaced with nonflammable materials.

Finally, the investigation ordered a complete overhaul of the quality control systems used to supervise construction. They had discovered an amazing complacency and negligence in both NASA and the contractors, not from greed or maliciousness but from simple overconfidence. As Borman said to Congress, "Quite frankly, we did not think, and this is a failing on my part and on everyone associated with us, we did not recognize the fact that we had the three essentials, an ignition source, extensive fuel and, of course, we knew we had the oxygen."

Borman now flew to Downey, California to act as the head of a NASA team helping North American Aviation redesign the spacecraft. George Low, the new head of the Apollo spacecraft program in Houston, had asked Borman to act as his "alter-ego," to make sure the redesign at North American was done correctly.

Borman spent the next year helping to incorporate the investigation's recommendations into a newly designed Apollo capsule.

## SOYUZ 1

Floating 125 miles above the earth, Cosmonaut Vladimir Komarov announced that "on the eve of the glorious historic event, the fiftieth anniversary of the Great October Socialist Revolution, I convey warm greetings to the peoples of our homeland who are blazing mankind's road to communism."[5] After more than two years, the Soviets had finally returned to space.

Komarov, who had piloted the October 1964 three man Voskhod mission, had blasted off early on the morning of April 23, 1967 in a brand new spacecraft, intended by the Soviets not only to take the first human to the moon, but to act as the foundation for their restructured long-term space program. With this spacecraft the Soviets hoped to build space stations and colonize the planets, establishing communism throughout the solar system.

Soyuz 1 was different from any previous spacecraft. Like the Gemini and Apollo capsules, it carried its retro-rockets and fuel tanks in a special equipment module attached to its rear. Unlike Gemini and Apollo, Soyuz actually had two separate habitable sections for the cosmonauts. The front section, called the orbital module and for use only in space, had a docking port for linking up with another Soyuz.

Attached to the orbital module was the crew module, bell-shaped with a flat heat shield at its base. When the time came to return to earth the cosmonauts would eject the round orbital module and return in the crew module.

Other innovations included the use of solar panels to provide energy rather than the fuel cells and batteries that American spacecraft used. When Soyuz reached orbit, the two winglike panels would unfold on each side of the equipment section.

Komarov's mission called for him to spend a day in orbit checking out the new Soyuz craft. Then another Soyuz with three cosmonauts aboard would blast off, chase him much as Gemini 6 had done with Gemini 7, and then dock. Because the docking port in this first Soyuz design contained no tunnel for crew transfers, two cosmonauts from Soyuz 2 would open the hatch and space walk across to Komarov's Soyuz 1. The three would then undock their craft from Soyuz 2 and return to earth, leaving the fourth cosmonaut to spend several more days in space.

Though Soviet space flights were no longer scheduled merely to upstage the West (as they had been under Khrushchev), the desire to score political points nonetheless played a part in Komarov's flight. Despite serious life-threatening failures on all three previous unmanned Soyuz test flights (one exploded on the launchpad and the other two had attitude control failures causing serious damage during reentry), Brezhnev pressured his engineers to fly the fourth Soyuz flight manned. With the Apollo investigation just completed, he wanted a political triumph to demonstrate the superiority of Soviet technology.[6]

Unfortunately, things began going wrong as soon as Komarov reached orbit. Only one of the two solar panels unfolded, cutting his electrical power in half. His main shortwave radio failed, forcing him to use his backup radio.

Then the system for keeping the spacecraft properly oriented began to fail. Unlike American space capsules, the Soyuz craft had been designed so that it could be automatically piloted from the ground. In fact, Soviet cosmonauts were intended to be merely backup observers, not pilots.

Hence, when the automatic control system broke down and Komarov was forced to manually pilot his craft, he was using thruster controls that were difficult to use and functioned only sporadically.

By now mission control had canceled the second Soyuz launch and were desperately trying to figure out how to get Komarov back to earth alive. Unfortunately, their ground-to-capsule communications network was not very complete, and from the seventh to thirteenth orbits, more than nine hours, they had no way to talk to Komarov. "Try to get some sleep," they told him as his tumbling capsule moved out of range.

When they finally regained contact nine hours later, Komarov reported an almost completely out of control spacecraft. Recognizing that he might not survive reentry, the flight director had Komarov's wife Valentina brought to mission control so she and Komarov could have a last few precious minutes of private conversation.

Finally, on the seventeenth orbit Komarov manually fired the retrorockets. Now he had to hold the bucking spacecraft with the heat shield in front, protecting him for the searing fire of reentry.

Miraculously, he succeeded. The crew module's drogue chute deployed as planned, and it looked like Komarov would survive.

But now the spacecraft's main parachute failed to release. In a last ditch attempt to save his life, Komarov deployed his reserve chute, only to have this tangle with the still attached drogue. Plummeting earthward at more than four hundred miles an hour, the Soyuz 1 craft smashed into the soil of Russia, south of the Ural mountains.

Komarov was killed instantly.

Decades would pass before the details of this tragedy became public knowledge, both in and out of the Soviet Union. Soviet officials merely announced that Komorov had "completed the full and complex program of testing the systems of Soyuz 1" during his one day flight, that he had then been "asked to stop flying and land," and that his death occurred when his "parachute system did not work."[7]

Hidden behind their tight-lipped secrecy, however, was an inconsolable sorrow and regret. Komorov's death was even more devastating to the Soviets than the Apollo launchpad fire was to the Americans. Never again would they fly a cosmonaut on a spacecraft that hadn't been tested thoroughly. Never again would they let their desire to be first allow them to take such chances.

Unbeknownst to the Americans, this newfound Soviet caution would have significant consequences on the outcome of the race to the moon.


COLUMBIA


On April 9th, 1968, an overflow crowd gathered in a small chapel on the Columbia University campus in New York City and listened as Reverend D. Moran Weston read excerpts from the works of Martin Luther King, Jr.

Five days earlier King had been standing on a motel balcony in Memphis, Tennessee, chatting with some friends, when he was gunned down by an assassin's bullet. Almost immediately riots and looting broke out in New York, Washington, Chicago, Detroit, Boston and several other cities. In Washington more than six hundred fires were set, with entire blocks burning to the ground. Twelve thousand federal troops were called out to patrol the city. In Chicago, the riots killed nine people and injured three hundred, requiring five thousand guardsmen to stop the violence.[8]

The Columbia University community now gathered to mourn the death of the civil rights leader. Weston finished speaking, and then asked everyone to hold hands and sing "We Shall Overcome."

Then David Truman, university vice-president, stood up to say his own eulogy for Reverend King. Before he could reach the podium a twenty-year-old young man with scruffy black hair stepped up onto the stage. The student's name was Mark Rudd.

"Columbia's administration is morally corrupt, unjust and indulges in racist policies," Rudd shouted. He then reeled off a series of charges about the university's anti-union policy and its construction of a gym in a public park adjacent to the campus. "If we really want to honor this man's memory then we ought to stand together against this racist gym." Rudd then marched from the room, followed by about three dozen followers.[9]

The tensions at Columbia had been fed and nurtured by a series of increasingly violent events in the last three years, both in America and in the jungles of Vietnam. Since 1965, riots in urban black neighborhoods had become an almost annual occurrence, with looting and bloodshed in New York, Chicago, and Cleveland in 1966 and Buffalo, Boston, Cincinnati, Detroit and Newark in 1967. Most required the National Guard to enforce peace.

In Vietnam, the Johnson Administration had increased the number of American troops to just under 550,000 men.[10] In the three years since the Tonkin Gulf resolution bombing missions over North Vietnam had become a daily routine. By the beginning of 1968, the war had claimed the lives of 16,000 Americans.[11]

Then in late January, 1968, the war exploded. During the Lunar New Year holiday of Tet, the Vietcong unleashed their biggest offensive. Beginning in the central areas of South Vietnam, the assault soon spread across the entire country, from the northern city of Hue to parts of the Mekong delta, south of Saigon. More than half the country's provincial capitals were attacked. In Saigon guerrillas stormed the U.S. embassy, setting off mines and occupying part of the embassy grounds for over six hours before being killed. For a short while the Vietcong managed to close all roads into Saigon, as well as forcing the shutdown of the city's airport. Soon, parts of the city were evacuated so that U.S. combat jets could bomb Vietcong-held neighborhoods.

Though the North Vietnamese were driven back, unable to hold any of their gains and losing somewhere between 10,000 to 20,000 soldiers, the offensive succeeded in planting in the United States widespread doubt of the country's ability to win the war as well as of the legitimacy of the South Vietnam government. In driving the North Vietnamese out of Saigon, Americans were witness to a public execution. The executioner, a South Vietnamese general, claimed that because the captured man had a handgun he was a Vietcong terrorist. With news cameras rolling, the general pulled out his pistol, put it to the prisoner's head, and shot him.

Within weeks politicians from both parties, including Eugene McCarthy, Edward Kennedy, Robert Kennedy, and Jacob Javits, were calling for an end to American involvement in Vietnam.[12] At the same time the two leading Republican candidates for President, Richard Nixon and George Romney, renewed their attacks on Johnson's policies.[13]

By April 1968, when Mark Rudd stepped up to the podium to condemn the Columbia University administration, U.S. casualties in Vietnam had risen to almost 22,000 deaths and only eight days earlier had cost Lyndon Johnson the presidency.[14] Johnson, having never clearly defined the goals of that war and faced with a rising storm of protest within his own party, had bowed out of the race for reelection.

At Columbia University, the fury over this unwanted, badly-fought war barely simmered below the surface. At the center of that anger was Mark Rudd and his followers.

Rudd was the head of the Columbia University chapter of the Students for a Democratic Society (S.D.S.). With about 30,000 members nationwide, the S.D.S. had for several years helped organize many of the earliest, most visible antiwar protests, such as the November 1968 rally in Washington.*

---

* See page 125

Many in the S.D.S. leadership were hostile to the United States and capitalism, often expressing themselves in words reminiscent of Soviet propaganda.[15] "What we are witnessing and participating in is a revolt of the trainees of the new working class against oppressive conditions of capitalism," wrote S.D.S. vice-president Carl Davidson.[16] Or as local Columbia University chapter officer Anthony Papert said, "A university controlled by imperialists is not going to allow these changes. So the practical answer is we'll have to take it over."[17]

For Rudd, the specific reasons for protest, "to end university complicity with the [Vietnam] war," were perhaps less important than the protest itself.[18] In 1970 Roger Kahn, in his book about the Columbia University protests, described Mark Rudd as

> A curiously appealing young man, except when he is possessed by a vulgarity or hostility or arrogance. He speaks earnestly and forcefully about a new order. He wants to see mankind freed from toil. How? He is not certain, and he does not take suggestions well. When someone corrects Rudd . . . his rhetoric grows simple. "Aw, fuck off."[19]

In October Rudd had written what he called a "coherent strategy" for "radicalizing" the Columbia student body. His game plan called for a step-by-step escalation of the conflict, beginning with position papers and leading to petitions, "harassment of [ROTC] instructors," demonstrations, and finally "a general student strike."[20]

Soon to join the Vietnam War and the "oppressive conditions of capitalism" as issues of protest was the gymnasium that Columbia was building in Morningside Park. Seven years earlier the university and New York City had made a deal: the university would be given a plot of land in the park in exchange for allowing community use of the gym.

By 1965, however, a number of politicians and local Harlem community leaders were questioning whether the city should have been leasing public park land to a private organization. By the spring of 1968 the S.D.S. had joined them, demanding that the gym's construction be stopped.

On Monday, April 23rd, approximately five hundred students, some from the S.D.S., some from the Students' Afro-American Society, and some merely curious bystanders, gathered in the center of the campus. There at the bottom of the grand steps leading up to Low Library they chanted and made speeches, condemning the gym as Rudd had done at the King memorial. Then about two hundred demonstrators proceeded to the gym construction site, where several attempted to tear down the chainlink fence that surrounded it. Three policemen intervened and a scuffle ensued. One student was arrested when a policeman was knocked down and kicked.

In retaliation, the protestors returned to the campus to grab their own "hostage." They moved *en masse* to the ground floor of nearby Hamilton Hall, where Rudd told a dean and two other faculty members: "We're going to keep you here."[21]

Though the "hostages" were released the next day, by Thursday, five buildings were occupied and the campus was shut down. When one hundred fifty students seized Low Library, they broke into the locked building by smashing a window and injuring the security guard. Once inside they rummaged through desks and rifled the files of the university president.[22]

Soon politicians and outsiders were inserting themselves into the conflict. Roy Innis from the Congress of Racial Equality (CORE) showed up to praise the demonstrators. "I'm proud of these kids. They've got the dean in what you might call *an extended dialogue.*"[23] State Senator Basil Patterson arrived to negotiate for the students. When he could get no satisfaction he said, "I know Harlem . . . settle this by nightfall."[24] Tom Hayden, one of the early founders of the S.D.S., made an appearance and suggested that if the police moved in, the protesters should make use of the university's collection of priceless oriental vases as a defense. "We take these pieces and put 'em out on the ledges. First time a cop takes a step toward us, we shove off a Ming vase."[25]

The university's administration and faculty stood by, unable to make up their minds exactly what they should do or where they should stand on any of the issues. The administration feared riots in nearby Harlem, and thus waffled on doing anything. The faculty wanted the protest to end, but opposed police intervention. Many, wearing white armbands to identify themselves, stood in front of the occupied buildings, and like school crossing guards, ushered the protesters in and out as well as guiding both demonstrators and counter-demonstrators into neat and proper lines.

Finally, on Monday night, the administration decided to call in the police. In the early predawn hours of Tuesday a force of about a thousand policemen was sent in to retake Columbia University. In front of each occupied building about a hundred officers arrayed themselves, with the officer in charge announcing by bullhorn that "On behalf of the trustees of Columbia University . . . I have been authorized to order you not to remain and you are hereby ordered to remove yourselves forthwith." At two buildings, underground tunnels allowed additional police officers to slip inside from below.

Only one building was cleared peaceably. Hamilton Hall had been the first building occupied. After the first day the black protesters ordered the whites to leave so they could take sole control of the building. Now the blacks calmly told the policeman in charge that they would not resist arrest, but that they would only leave if arrested. All eighty-three were handcuffed and quietly led to vans and taken away.

Everywhere else was violence. Students, both inside the buildings and in crowds outside, screamed "Pigs!" "Fascists!" and "Motherfuckers!" at the police. At some buildings the students and facility outside linked arms and tried to block police access to the buildings. Inside, protesters also linked arms, refusing to stand and leave when ordered to do so.[26]

The cops retaliated with force, using batons and nightsticks. Pushing their way through the crowds, they dragged and shoved students from the buildings and into waiting police vans. Before long, the pushing and shoving changed to kicking and beating. Some cops cursed the students, shouting, "Commies!" "Bums!" and "Motherfuckers!" Soon students and teachers alike (even those who had decided not to resist) were assaulted. In front of one building the police formed a gauntlet, and forced every captured protester to run through it, bludgeoning them unmercifully as they passed. By night's end over one hundred people were injured, and over seven hundred were arrested.[27]

This was just the beginning of the protests and violence. On May 17th 117 persons were arrested when about 1,000 people, including both students and some local politicians, occupied a university-owned tenement on 114th Street.[28] On May 21st, sixty-eight persons were injured, including seventeen policemen, and another 177 arrested when students once again occupied Hamilton Hall.

This time it was the students who went wild, smashing windows and doors, setting fires, and throwing rocks, bottles, and bricks at police.[29]

What made the Columbia University demonstration so shocking was that it involved the nation's so-called elites — its upper middle-class students, its Ivy League intellectuals, and its law enforcement officials.

The students had been expected to spend their time learning about the world so that they could some day run it. Here, the protesting students seemed less interested in learning than in destroying the society around them.

The teachers had been expected to understand these complex issues, and to wield wisdom in the name of justice. Here, they merely stood by, helpless, unwilling to take any stand.

The police had been expected to firmly but justly enforce the law. Here, out of resentment and anger at the anti-American beliefs of some protesters, they violently broke it.

Expected to lead the country away from violence and irrational behavior, the participants at Columbia all reveled in it.

Possibly the most disturbing aspect of these protests was that, while the police and the school were justifiably condemned for their improprieties, the demonstrators were in the years to follow portrayed as noble heroes. As Jeff Kaplow, a thirty-year-old assistant professor at Columbia noted soon after, "I'm very sympathetic to [the] S.D.S. and I don't deplore the taking of the buildings. It's a silly piety to deplore it. It was done in a situation where all other remedies had failed."[30]

This is not to say that all protesters condoned this violence. Many people of good will participated in many peaceful Vietnam protests in the ensuing months. On April 24th, for example, with Columbia's buildings still occupied, almost 200,000 college students throughout the New York metropolitan area gathered in Central Park to peacefully protest the Vietnam War.[31]

Nevertheless, the aftermath of the Columbia protests could be seen almost immediately. Just three weeks later a group of forty students seized the registrar's office at Brooklyn College, occupying it for sixteen hours, demanding that the college guarantee the admission of 1,000 more blacks in the coming fall semester.[32] And this was only the beginning, as similar protests soon broke out in hundreds of campuses across the country.

American society had seen the arrival of a violent protest movement that in years to come would tear at the social order, attacking and changing every assumption about the country.

## SATURN 5

Twelve hours before Martin Luther King was shot in Memphis, the sun rose bright and clear at Cape Kennedy. At pad 39A, the countdown of the second unmanned test launch of the Saturn 5 rocket reached its finish. At 7 AM, the engines fired and, like a huge lumbering titan, the rocket began its slow climb skyward.

Though less than twenty months remained before the arrival of Kennedy's self-imposed deadline, NASA had only tested the complete package of this massive machine once before. In its first launch, on November 9th, 1967, the Saturn rocket had taken a test command module to a height of over 11,000 miles, at which point the service module's engines had driven the module back into the earth's atmosphere at almost 25,000 miles per hour. This accelerated return was intended to simulate the return of a spacecraft from lunar orbit. Its parachutes unfurling, the command module had landed safely in the Pacific, less than ten miles from the *U.S.S. Bennington*. It had been a perfect flight, putting a strong positive spin on what had been a sad year at NASA.

This second launch, officially named Apollo 6, would be a repeat of that first mission, giving the ground crew a bit more experience for the first manned mission now scheduled for October. It would also reaffirm that the Saturn 5 and all its components were ready and able to safely put human beings into space.

At one second past 7 AM, just as planned, Apollo 6 lifted off. Little else went as planned. Just over two minutes after launch, the engineers at the Marshall Space Flight Center in Huntsville, Alabama began recording powerful oscillations in the thrust of the rocket. Like a badly-tuned car, the rocket was actually *bouncing* five or six times a second as it rose into the air, the rocket's thrust fluctuating up and down across a wide range of acceleration. For ten seconds these pulses surged along the length of the rocket — pulses so violent that a large outside panel was ripped away.

Then, just as the center engine of the first stage shut down on schedule, the pulses stopped. Fifteen seconds later the other four engines turned off, and the first stage was jettisoned on schedule. Now the five engines of stage two ignited, propelling the rocket forward with a million pounds of thrust. About four minutes into the burn, however, one of these engines began to have trouble. Then it cut off inexplicably, followed almost immediately by the cut-off of a second engine. Normally, the loss of two engines required the ground controllers to immediately abort the flight, but somehow the rocket was managing to balance the thrust of the remaining three engines. Making a split-second decision, flight director Charlesworth and controller Bob Wolf allowed the rocket to keep flying, burning the remaining three engines of the second stage for almost five more minutes, fifty-nine seconds longer than originally planned and until its fuel tanks were dry.[33]

When the rocket's third stage engine (the S4B) kicked in, the ground controllers extended its burn as well in order to compensate for the lost thrust. As a result they managed to get the spacecraft into a wobbly orbit, 110 miles at its lowest altitude and 228 miles at its highest.

After allowing the ship to make two circuits of the earth, the controllers now tried to simulate a command module's return from the moon. By firing the S4B engine they would push the command module *down* into the atmosphere, increasing its reentry speed to over 24,000 miles per hour.

Unfortunately, the S4B engine refused to re-ignite. Now the controllers were forced to fire the S.P.S. engine instead, extending its burn to almost seven minutes — two and a half minutes longer than planned — in an attempt to reach the desired speed. Upon hitting the atmosphere, the command module was flying only 22,000 miles an hour, and it splashed down fifty miles from its planned landing point in the Pacific.

For NASA, the assassination of Martin Luther King acted to divert attention from these problems. The failure of Apollo 6 to perform as expected was hardly noticed by the general public.

Yet, with only a year and eight months before the end of the decade, not only had three engines of the Saturn 5 not worked according to plan, but one whole outside panel had been torn from the rocket, and the oscillations during launch would have been strong enough to injure any astronauts on board at the time. "It was a fascinating flight," controller Jay Greene said tersely many years later.[34]

If NASA was going to send men to the moon in the next twenty months, it would have to find out very quickly what had caused these problems, and solve them just as fast.

And the pressure was building. The Soviet space program was coming back to life. On March 2nd, one month before the unsuccessful Apollo 6 mission, the Soviets launched Zond 4. Since the death of Komarov eleven months earlier, the Soviet space program had established new flight guidelines. No manned space flight would take place until an identical unmanned robot spacecraft had successfully accomplished an identical mission.

Zond 4, of a similar shape and configuration to the Soyuz crew and service modules, was intended to prove that this craft could safely return a human from lunar space. The capsule was lifted to an distance of about 205,000 miles, and then, just as had been planned for Apollo 6, dropped back to earth in order to test high velocity reentry procedures. While the launch was successful, Zond reentered the atmosphere at too steep an angle. At six miles altitude the Soviets blew the craft up, preventing it from crashing into west Africa, where they feared Western technicians might recover it.

Then, three days after the unsuccessful launch of Apollo 6, Luna 14 was launched, and after a three-day journey entered lunar orbit. According to Tass's cryptic description, the orbiter was there to "conduct further scientific studies of the near-lunar space."[35] Speculation abounded, however, that this was a test flight of a Soviet manned lunar vehicle.

The Soviets then topped this event one week later with the launch of Cosmos 213 and Cosmos 214. These Soyuz unmanned test craft successfully achieved the automatic docking maneuver that Komarov had been unable to attempt when his spacecraft failed one year earlier.

Shortly thereafter, Frank Borman met in Houston with George Low, manager of the Apollo Spacecraft Program. Borman, not one to mince words, described how well the capsule redesign was going. In his mind, the problems that had caused the launchpad fire more than a year earlier were solved. When construction on the first few Apollo capsules was finished in the next few months, they would be ready to take men back into space.

As soon as the meeting ended, Low decided it was time to assemble a team to see if a flight to the moon was possible before the end of the year. He wrote a memo on the subject, and told his secretary to consider that memo "007," which meant that once it was read by his superiors at NASA, she was to destroy it, not even keeping a copy for Low's files.[36]

Despite the death of three astronauts only fifteen months before, despite the failure of Apollo 6, and despite the lack of any manned test flights of a Saturn rocket, George Low believed the time had come to send a man to the moon. And he was convinced that they could be ready to do it in less than eight months.

## CHAPTER NINE

# "THERE'S A BEAUTIFUL EARTH OUT THERE."

"YOU ARE GO FOR REV TWO. All systems are go." After an hour's careful review of the S.P.S. engine's performance, the evening's flight director, Milton Windler, had okayed a second lunar orbit for Apollo 8.

Five-forty in the morning, Tuesday, Christmas Eve 1968. Though the astronauts had catnapped during the early evening hours Monday night, none had had a full night's sleep since Sunday, and Anders hadn't even had that.

Now that they were in lunar orbit, however, there was no time for sleep. Even if everything went flawlessly, they would only be there for another eighteen hours, and they had much to accomplish in that short time.

As the sun rose on the morning of Christmas Eve in Houston, Apollo 8 slipped behind the moon for the second time.

\* \* \*

On earth, the three families sat tensely in their homes, awaiting the first lunar telecast scheduled for 6:30 AM. Though the men had made lunar orbit safely, this was now the period of greatest risk.

After the spacecraft had entered lunar orbit Valerie Anders had tried to go back to sleep. Normally a sound sleeper, she lay in bed alone, listening to the squawk box. Once again the men's voices, rather than lulling her to sleep, kept her awake. As dawn approached and her kids began to stir she gave it up and climbed out of bed.

Marilyn Lovell had also tried to sleep, without success. She dozed in her bed, lulled by the drone of the squawk box jargon. By dawn, however, she was back in front of the television, awaiting that first telecast.

Susan Borman never even tried to sleep. She knew she wouldn't sleep until the men left lunar orbit. She sat in her kitchen, intently watching the television and listening to the squawk box.

Hidden behind the moon, the astronauts hurried to set up their 16mm movie camera. With only five minutes before they moved into daylight, the flight plan said that the camera should be running automatically as they glided over the moon's stark, colorless surface.

As the sun came up, the three men focused on taking pictures. Anders, in charge of photography, kept moving back and forth between the few good windows, alternating from still to movie cameras, while also directing the other two men in what to photograph. The work was made somewhat difficult in that three of their five windows were practically unusable, covered with what Anders later described "as purplish smears, as if a service station attendant had attempted to clean a windshield using an oily rag."[1]

Anders' photography was guided by some carefully detailed reconnaissance maps, produced from photos taken by a number of unmanned lunar orbiters, both American and Russian.[*] These charts not only told him what parts of the lunar surface he was flying over at every moment of the flight, they also indicated what camera shutter speeds and f-stops he should use.

While the moon's major nearside landmarks were named, the far side's features were mostly nameless. To make his job easier as well as honor a number

---

[*] While there had been no cooperation between the two space programs, the Soviets in 1960 had published an atlas of the moon's far side based on photos taken by their Luna 3 probe. This data helped supplement the surveys taken by NASA lunar orbiters

of people, Bill had taken the explorer's prerogative and put names to these craters and mountains. Three craters he named Grissom, White, and Chaffee. Others he named Kraft, Slayton, Low, Von Braun, and Shepard. Some he named Mercury, Apollo, Texas, and Washington.

And shrewdly, he picked three distinct craters just on the edge of the farside's horizon, just out of sight of earth, to name Borman, Lovell, and Anders. Though the earth could never see these features directly, from his position in orbit Bill would be able to relay a television picture of them back to earth.

Soon it was time for that first lunar telecast. The flight plan called for a televised press conference the moment they came out from behind the moon. After several minutes' effort to set up the television camera, the astronauts switched it on just as they regained contact with the earth. The picture showed a bright lunar surface silhouetted by the window frame.

This was mostly Bill Anders's show. For the next eleven minutes he shifted the TV camera between two windows, describing to the world the landmarks that were slowly drifting past. Lovell studied the maps to help him identify the less well-known features, while Borman said nothing as he steered, using the hand controls to keep the spacecraft's nose pointed down so that the surface was visible in the windows.

Ever so often Anders would add other details about the surface. "The color of the moon looks a very whitish gray like dirty beach sand, and with lots of footprints in it."

Once or twice Lovell added his own impression. "Don't these new craters look like pickaxes striking concrete, leaving a lot of fine haze dust?"

Mostly, however, Anders let the picture speak for itself. He wanted the people on earth to experience what it was like to orbit the moon.

Valerie watched and was disappointed at how bleak and splattered the moon looked. *No jagged edges,* she thought.

Marilyn Lovell watched with growing wonder. She looked at that bleak, beaten lunar surface and thought, *It's so vast; it's so empty.* There was no life there. Nothing at all.

Susan couldn't care less what the moon looked like. She just wanted Frank to succeed. Once again she closed her eyes and listened to their voices, trying to imagine herself in the capsule with him. She couldn't comprehend

Note the rilles inside and cutting across the largest crater.

it. Frank was as close to the moon as she was to Jim and Margaret Elkins in Houston. The distance and remoteness of it only fed her fear and pessimism.

She got up and left the house, keeping a beauty parlor appointment she had made previously. She'd be damned if she would allow her fears to conquer her.

\* \* \*

With the show over, Borman and Anders prepared for the second orbital burn, scheduled for when they were behind the moon. This would lower the high point of their orbit from 194 miles to seventy, thereby circularizing it.

Eroded mountain. Impact craters.

Lovell took the mike and continued the description of the lunar surface, reporting: "The view at this altitude is tremendous. There is no trouble picking out features we learned on the map . . . I wish we had the TV still going." In lunar orbit Lovell's job was to study the lunar surface from the perspective of navigation. Could the navigational charts, created from Lunar Orbiter photos, be used to find one's location on the lunar surface? Were the prime landing spots for future Apollo missions acceptable? Or were there any unidentified obstacles that might cause problems? And could an astronaut see those obstacles and pilot his way through them?

He described such features as the Sea of Tranquility, the crater Taruntius, and the mountains that skirted both. "The mountain range has got more contrast, because of the sun angle," he noted. Then he inexplicably said, "I can see the initial point right now, Mount Marilyn."[2]

Though Mike Collins on the ground responded, "Roger," he had no idea what Lovell was talking about. His knowledge of the area around the Sea of Tranquility included no such mountain.

Until Lovell had arrived that had been true. The ancient astronomers had never named Mt. Marilyn with their earthbound telescopes. NASA hadn't given it a name when their unmanned probes had photographed it. And neither Borman nor Anders had noticed it when they had studied the maps prior to launch.

Just as he had promised, Jim Lovell had brought his wife with him to the moon. Even if some bureaucrat in some scientific institution might not consider the name Mt. Marilyn official, it was now that mountain's name to him.*

Much like Anders, Lovell had chosen shrewdly. Placed on the edge of the Sea of Tranquillity just east of the Apollo 11 lunar landing site, Mt. Marilyn would be on all the maps used by Armstrong and Aldrin when they made their historic approach six months later. In fact, they would fly right over it as they came in for a landing.

The astronauts slipped behind the moon for the third time. As they had on the first orbit, Anders and Borman worked their way down a long checklist, making sure everything was right for the engine burn, while Lovell programmed the computer. Thirty minutes later the S.P.S. engine fired for the third time, burning for twenty-one seconds and lowering their orbit. It would now take them an hour and forty minutes to circle the moon instead of two hours.

---

* For what seem to be political and to this writer somewhat petty reasons, the scientific organization which has taken upon itself the task of officially naming astronomical objects, the International Astronomical Union (I.A.U.), rejected all of the names chosen by both Anders and Lovell. Instead, the I.A.U. named three small craters after the Apollo 8 astronauts in a part of the moon the astronauts never saw.

A map showing the location of both Bill Anders's named craters as well as Lovell's Mt. Marilyn is included in this book as my statement of disagreement with the I.A.U. It seems to me that the explorer, the person who risked his life to discover a new world (rather than some bureaucrat on earth) should always be given the prerogative of choosing the names.

Regardless, it will be up to posterity to decide. I include the maps here for the record.

This is the Lunar Orbiter photo that Jim Lovell gave to Marilyn
two nights before launch. Mt. Marilyn is the triangular-shaped mountain
at the bottom of the photograph just to the right of center. It is located slightly
north of the lunar equator. The crater Secchi is to the east, just out of frame.
Tranquility Base, also near the equator, is about 150 miles to the west.
See endsheet map.

The pace was unrelenting. Even as the men finished this burn they immediately resumed their photography and reconnaissance of the barren lunar surface. And when they came out from behind the moon, Borman and Lovell once again went through the tedious but critical routine of taking down new numbers for the computer, while Anders reviewed the status of the fuel cells and life support systems with the ground.

Just before 10 AM, Frank Borman decided it was time to do something slightly less technical. He asked if Rod Rose was around.

Mike Collins told Borman that "Rod Rose is sitting up in the viewing room. He can hear what you say."

A month earlier, just before Borman had left for Florida, he attended Sunday church services at his local parish, St. Christopher's Episcopal Church in

League City, Texas. As one of the church's lay readers who routinely read aloud to the congregation during church services, Borman had planned to participate in the Christmas day services. When NASA made the sudden decision to send Apollo 8 to the moon, he found he had to explain why he couldn't be there. "We kidded Frank about going to such lengths — all the way to the moon — to get out of . . . services," said his reverend, James Buckner.[3]

The conversation soon turned serious. Borman really wanted to participate in that Christmas Day service, but didn't have any idea what he could do. Fellow parishioner Rod Rose, an engineer at mission control, offered a solution. He would put together a short prayer that Borman could read from orbit, tape Borman's recitation, and then play the tape back at church. For Borman, the practical test pilot, this plan was perfect. Rose cobbled together a prayer from a number of verses in the Bible, and went over it with Borman until both were happy.

Now, Borman waited until Lovell and Anders finished passing some new data to the ground. Then he began, a little self-consciously. "This is to Rod Rose and the people at St. Christopher's, actually to people everywhere." Borman took a breath. "Give us, O God, the vision which can see thy love in the world, in spite of human failure. Give us the faith to trust the goodness in spite of our ignorance and weakness. Give us the knowledge that we may continue to pray with understanding hearts, and show us what each one of us can do to set forth the coming of the day of universal peace. Amen."

"Amen," Mike Collins echoed softly.

Now Borman sheepishly added, "I was supposed to lay-read tonight, but I couldn't quite make it."

"Roger," said Collins. "I think they understand."

Susan, already back from the beauty parlor, heard the prayer and felt only relief that one more possible disaster had passed. She had become so pessimistic that even so simple a thing as Frank reading a prayer aloud seemed a miracle.

In the spacecraft the work went on. As soon as Borman finished reading his prayer, Collins asked him when they had last chlorinated their water. The astronauts' drinking water was produced as a byproduct of the generation of electricity by their fuel-cells. This water was transferred into a tank, and periodically the crew manually injected chlorine into it to keep the water free from bacteria. Borman confirmed that they had done so ninety minutes earlier. "Jim spilled a little, and it smelled like a bucket of chlorine in here."

Ten minutes later they slipped behind the moon for the fourth time. It was now midmorning, Tuesday. Except for those short naps twelve hours earlier, they had all been awake since early Monday morning.

But it wasn't yet time to sleep. Now Anders's photography became the prime mission focus. Though Borman's word was law when it came to running the overall mission, Anders usually supervised the taking of photographs. If someone else wanted to take a picture, they generally cleared it with him first to make sure it fit into the mission's objectives. Mostly, however, they left the photography to him.

As Anders worked he logged his subject by announcing what he was doing out loud, recording the information on the onboard tape recorder. "Okay, definitely frame 114, target of opportunity, extremely fresh impact crater. Time 75:39:30." For forty minutes, the other two men kept quiet, letting Anders do his job. Borman would continually adjust the spacecraft's attitude so that its nose always pointed downward, thereby keeping the lunar surface visible in the least obstructed windows.

Just before they came out from behind the moon, Anders finally stopped. Borman needed to roll the spacecraft so that Lovell could do a navigational sighting, and he and Lovell had been waiting patiently for Anders to finish. With only seconds before they reacquired earth signal Anders gave the okay, and Borman started the roll.

To do this Borman gazed out his window, using the moon's horizon as a reference point. Suddenly he noticed a blue-and-white fuzzy arch edging from behind the moon's sharp horizon line of dreary gray. This growing round patch was the only color in a black-and-white universe. "Oh my God!" Borman cried out. "Look at that picture over there! Here's the earth coming up!" He stared at the earthrise in wonder. "Wow, is that pretty."

Lovell gaped. "Oh man, that's great!"

Suddenly, they all realized that they had to get a picture. Of all the objectives NASA had set before launch, no one had thought of photographing the earth from lunar orbit. Borman grabbed the nearby floating camera that

Anders had been using and snapped a picture, only to have Bill joke, "Hey don't take that, it's not scheduled."

They all laughed. Borman handed the camera to Anders and looked out the window again. "Gee," Borman sighed. The earth was so beautiful, and so far away.

Anders also wanted to get a picture but the camera Borman had given him was loaded with black-and-white film. "Hand me that roll of color quick," he said to Lovell, who was closest to the correct storage locker. For a few moments there was panic as Lovell scrambled to get the film and Anders struggled to load it. Then they jostled for position at the window.

Anders: "Let me get it out this window. It's a lot clearer."

Lovell: "Bill, I got it framed — it's very clear right here."

Outside, the half-full earth had now risen several degrees "above" the moon's horizon. It glistened with a blue-white gleam against that jet-black sky, the moon's dead surface a stark contrast.

As Anders framed the shot Lovell hung over his shoulder, almost taking the camera from him in his desire to make sure the picture was taken. Borman had to tell him to calm down.

Lovell was entranced. "Oh, that's a beautiful shot." He asked Anders to take a number of pictures, varying the exposure.

Anders nodded. "I did. I took two of them."

"You sure we got it now?"

"Yes." He looked at Lovell dryly. "It'll come up again, I think."

The three men stared at their home planet as it drifted slowly into the sky. They had just witnessed the first earthrise ever seen by any human being.*

---

* For years there has been confusion over who took the earthrise picture that was seen on millions of stamps and magazine covers soon after the astronauts' return. Borman always claimed that he took it, and anyone who knew Frank Borman (including Bill Anders and Jim Lovell) knew he wouldn't make this claim unless he believed it to be true. Yet, the transcripts clearly show that Bill Anders took the color picture familiar to us all. For years, no one could quite understand why Borman appeared to be trying to claim credit for something he hadn't done — an action completely uncharacteristic of him.

The solution to the mystery is that more than one picture was taken. Borman took the first earthrise shot ever taken, but his black-and-white photograph on a different roll of film has been ignored all these years because Anders's later but prettier color shot of the same earthrise was available.

Because Anders so much wanted to get credit for his photograph, Lovell has spent a great deal of time needling his good friend about it in public. For example, Lovell simply refuses to state in public who took the picture, despite knowing that Anders took it. When I asked him why he does this, he grinned and said, "It keeps us young and happy."

The first earthrise picture photographed by Frank Borman. The cloud formations prove this, as they are identical to Bill Anders's picture, taken a few minutes later.

It was just before noon in Houston. Mike Collins had not yet given the astronauts his daily news update, and he now asked Borman if they were interested in hearing it. Borman wasn't. "I'll give you a call," he said, wanting at that moment no distractions. He needed to get Houston to dump the onboard tapes so that they could be reused during the next revolution. He needed data for the Trans-Earth Injection burn, or T.E.I., the engine firing thirteen hours hence that would blast them out of lunar orbit and send them back to earth. And he once again needed an official okay that they were go for another lunar orbit.

Finally, ten minutes before they slipped behind the moon for the fifth time, Collins gave them the news. "Your TV program was a big success . . . It was carried live all over Europe, including even Moscow and East Berlin . . . San Diego welcomed home today the *Pueblo* crew in a big ceremony. They had a pretty rough time of it in the Korean prison . . . Christmas cease-fire is in effect in Vietnam, with only sporadic outbreaks of fighting."

Borman listened with half an ear. As he had said to Collins just five minutes earlier, "We're tired right now."

As Apollo 8 disappeared behind the moon for the fifth time, Borman relinquished the controls and went to sleep.

For the next two orbits, while Borman dozed, Lovell kept the spacecraft oriented downward while Anders took pictures. With the commander asleep, the other two men seemed to relax somewhat, chatting about the experience so far. "It doesn't seem like we've hardly been here that long, does it?" Anders asked at one point.

"It seems like I've been here forever," Lovell replied. Finally where he had wanted to be since he was a child, Lovell couldn't get enough of it.

Another time Lovell joked that neither of them was a scientist. "All those scientists are saying now, 'Oh, if we only had a geologist aboard!'"

Anders looked at the fogged-up windows. "[A geologist] couldn't see anything . . . Nothing but a big blur out there.

"You know," Anders added, "[the moon] really isn't anywhere near as interesting as I thought it was going to be. It's all beat up." Like his wife, Anders had expected the moon to look like all the classic science fiction paintings, sharp-edged mountains and razor-cut ridges delineated by harsh black and white shadows. Instead, the mountains looked rounded and eroded, as if they "had been sandblasted through the centuries."[4] And as did his wife, he found this disappointing.

In many ways, Bill Anders could be called the first scientist to fly in space. While most astronauts had graduate degrees, almost all had been trained in aeronautics, the science and engineering of flight. Anders, however, had earned a degree in nuclear engineering. This gave him a slighter wider background in research and the hard sciences. In addition, during his five years of waiting for a space flight he had gone on every geological field trip offered by NASA. He found that he was interested in studying the many

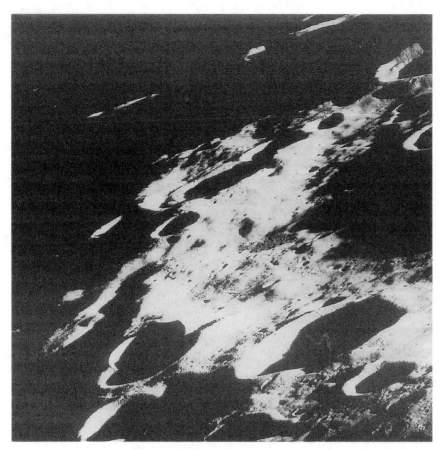

Lunar sunrise.

mysteries that haunted scientists about the moon's formation. Anders really wanted to help answer these questions.

The solution to one mystery seemed obvious from only seventy miles away. Scientists had debated for decades whether the numerous lunar craters were formed from volcanic activity or asteroid impacts, with most astronomers accepting volcanoes as the solution as recently as 1950. If of impact origin the solar system's entire formation history would have to be rethought, considering the enormous number of craters visible on the moon.[5] As Anders photographed the lunar surface he tried to describe what he saw. Though there was "some hint of possible volcanic . . . activity" in a few areas, he noted that almost every crater appeared to be of impact origin. He added that a manned landing on the far side of the moon would be difficult. "The backside looks like a sandpile my kids have been playing in for a long time. It's all beat up, no definition. Just a lot of bumps and holes."

Periodically Borman would open his eyes and though still half-asleep mumble a question about the time, the ship, the situation. The others would reassure him that all was fine, and he would drift back asleep.

At 4 PM, the spacecraft moved behind the moon for the seventh time. Lovell was still at the helm, and humming and singing aloud as he guided the ship through space.* Anders worked the cameras. Neither had slept.

Borman, however, was finally up, but he wasn't ready to return to work. He ate, used the "Waste Management System," and joked with the other two men.

At one moment Lovell looked out the window, and then at his crewmates. "Well, did you guys ever think that one Christmas you'd be orbiting the moon?"

Anders quipped, "Just hope we're not doing it on New Year's."

Lovell, who was usually the joker, didn't find this funny. "Hey, hey, don't talk like that, Bill. Think positive."

\* \* \*

Two hundred forty thousand miles away, Marilyn Lovell decided she needed to go to church. She had spent the day of Christmas Eve at home with friends, trying to fill the time as the spacecraft orbited the moon. Periodically she wandered back to her bedroom for some quiet. Other times she listened to the squawk box. And she smoked a lot.

Finally, by late afternoon the magnitude of the day's events were wearing on her. She wanted to pray, but she wanted to do it alone. She called Father Raish at St. John's to ask him if she could come to church, and he told her to come right over.

---

* Lovell sang aloud repeatedly throughout the mission. Neither he, Borman, nor Anders remember the tunes, however, and the transcripts do not say, stating merely that he was "(singing)."

As soon as Father Raish hung up the phone he ran to find the church organist, asking her to play the musical program planned for that evening's midnight mass service. Raish then went throughout the church and lit all the candles, making the place look like it would for that night's mass. He knew that the astronauts would be leaving lunar orbit just after midnight, which meant that Marilyn couldn't attend services. Nonetheless, he wanted her to have the experience.

When Marilyn entered the church a few minutes later she gasped. "It was one of the most beautiful sights that I can remember. And it was all for *me*."

She and Father Raish went up to the altar together, kneeled down, and prayed. For Marilyn, this was a profound moment in her life. "It meant so much to me that he did this." Even today it brings tears to her eyes to think of it.

After a short while it was time to leave. By now the sun had set, and as Marilyn drove home the sky was dark.

Because the Houston sky had been cloudy since the launch on Saturday, the astronaut wives had still not seen the moon. Even now Marilyn could see that the evening sky was mostly covered with clouds.

Then, just before she turned the car into Timber Cove, the clouds separated and she found herself staring at the faint arc of the crescent moon, gleaming in the evening sky. It was Christmas Eve, and around that distant globe circled three humans, one of whom was her husband. A shiver ran through her body.

She pulled up in front of her house and rejoined her children and friends for the last lunar telecast and the hoped-for exit from lunar orbit.

\* \* \*

Susan Borman had no religious outlet. Though she believed in God, her faith in NASA had disappeared the night Ed White died.

Remembering how Pat fought with NASA over Ed's funeral, Susan now decided that she wasn't going to be caught unprepared. Sometime that afternoon, even as she listened to Frank's voice on the squawk box, she sat down at the kitchen table and began to write out the memorial service she wanted for him.[6]

She wondered how she was going to live with his death. She then rationalized, *What a magnificent place to die!* She began to write the words, about how no one should be sad, that everyone should be comforted because Frank is still there, in orbit, for ever and ever. *That would be what Frank would want,* she thought.

The words she scribbled onto that piece of paper seemed to express for her how Frank's death in orbit would complete both their missions. He would have made the greatest possible sacrifice for his country, and she would stand before the world and tell them so.

Fifteen-year-old Ed wandered into the kitchen and saw her writing feverishly. He asked her what she was doing.

She explained, "Your father's memorial service. He might not come back."

From Ed's young perspective, this simply wasn't possible. He took the pen from her. "Just remember, Mom, Dad gets to choose the way he goes — you and I don't have that privilege."

Susan nodded. But she took the sheet with her words and carefully hid it under some clothes in a bureau drawer. She was convinced she would need it.

Shortly thereafter they got into their car, picked up Frank's parents, and drove into Houston for Christmas dinner with Jim and Margaret Elkins and their children. Here she would have a few quiet hours free from the zoo of journalists and television cameras. Here she would watch their last press conference from lunar orbit. And here she would find out what Frank and the others had decided to say to the world on Christmas Eve. Though she knew they had planned a special message for this telecast, she didn't know what it was.

\* \* \*

Borman stared out the window and was pleasantly surprised to see his second earthrise. He felt strangely detached from that tiny blue-white planet. Somehow, his entire existence in the blackness of space was now contained in their tiny capsule and its "environment of winking amber and red instrument lights."[7]

What those lights indicated, however, meant the difference between life and death, and the lack of sleep was finally beginning to interfere with how the three men were reading those dials. Even as Borman contemplated that slowly rising earth, both Anders and Lovell made separate errors inputing data into the computer. Borman listened, and suggested that it was "time to take a rest."

Lovell nodded, "Okay, just a minute." Neither he nor Anders wanted to go to sleep. There was too much to do, and how could the first humans orbiting the moon waste time sleeping?

Yet, as they regained contact with the earth on this seventh orbit, all three men were clearly slowing down. Sometimes they had to ask questions twice, and sometimes they didn't understand the answers.

Lovell to Borman: "How do you feel?"

"Fine. Why?"

"I was just curious."

"You tired?"

"Oh, I'm a little tired," Lovell nodded. "I guess we all are."

And yet, neither he nor Anders could bring themselves to take a break.

At 5:40 PM Borman told Houston that he wanted to scrub some of Lovell's duties on the next revolution, so he could get some rest. Then he asked Anders if he wanted to get some sleep as well. Anders said no.

Borman wasn't satisfied. He knew that on orbits nine and ten, leading up to the T.E.I. burn, they all had to be sharp and on the money. Their lives depended on getting that burn right.

While Lovell was already getting ready for bed, Bill Anders resisted. He didn't feel tired, and he still had a great deal of unfinished photography on his flight plan. "Hey, Frank, how about on this next pass you just point [the camera] down and turn the goddamn cameras on. Let them run automatically."

"Yes, we can do that." Borman really didn't want to prevent Anders from finishing his job.

At that same moment, Borman suddenly realized he had ruined a roll of film. He cursed, and then his instinct to make fast decisions kicked in. He was no longer going to negotiate. He knew Lovell and Anders needed rest, and as commander of the mission it was his responsibility to make sure they got it.

Also at that moment, Ken Mattingly radioed to confirm the tasks that Borman had scrubbed from Lovell's flight plan. Borman instantly responded,

"We're scrubbing everything. I'll stay up and point, keep the spacecraft vertical, and take some automatic pictures, but I want Jim and Bill to get some rest." He looked at the flight plan. "Unbelievable — the detail these guys study up. A very good try, but just completely unrealistic."

Anders still resisted. "I'm willing to try."

"No. You try it, and then we'll make another mistake, like entering instead of proceeding or screwing up like I did. I want you to get your ass in bed. Right now." Borman had had enough of this conversation. "Go to bed! Hurry up! I'm not kidding you, get to bed!"

Anders didn't move. For the next five minutes he hung there, gently offering suggestions to Borman about how to set the camera up. He knew that as commander Borman had the right and authority to order him to bed. Anders just didn't want to go to sleep. How often would he get a chance to orbit the moon?

At 6 PM they slipped behind the moon for the eighth time. Just before, Ken Mattingly radioed NASA's support of Borman. "We agree with all your flight plan changes. And have a beautiful backside. We'll see you the next time around."

Still Anders resisted. He kept trying to find a way around Borman's order. While he didn't directly disobey it, he also didn't follow it immediately either.

Borman understood this. He figured he could just wait Anders out. He answered his questions by telling him to go to sleep. "I think this is a closed issue . . . I don't want to talk about it . . . Shut up, go to sleep, both of you guys . . . You should see your eyes — get to bed . . . Don't worry about the exposure business, goddamn it, Anders, get to bed. Right now! Come on!"

Finally Anders began to lose the battle, not so much with Borman but with his own body. He had hardly slept since launch, and next to him Lovell was already sleeping. Anders started to doze, but fought it. He asked Borman if the cabin was cold. Borman said, more gently this time, "Well, you're tired. It's not cold."

Soon Borman sat in silence in the command seat, staring out at the stark lunar surface. Despite knowing that Anders especially resented this forced nap, Borman had never shirked from making the hard decision when he knew it was the right thing to do. And he wasn't going to start now.

For the next few hours, he sat and observed the lunar surface, periodically taking pictures. Below him passed what for eons humans had called the dark side of the moon, that unseen hemisphere whose face was always turned away. Now a human being was not only able to study it, he was flying a mere seventy miles above its surface.

At 7:30 PM he asked Ken Mattingly, "How is the weather doing down there, Ken?"

"Entirely beautiful. Loud and clear, and just right in temperature."

Borman wasn't really interested in the weather in Houston. "How about the recovery area?" he asked. He knew that once they fired the S.P.S. engine to send them back to earth, it would be very difficult to change their arrival time. If the weather in the Pacific turned bad, they would have to land anyway.

"That's looking real good."

"Very good," Borman said. He looked at the tiny cloud-covered planet in his window. It would be very nice to be back there.

Ken Mattingly felt he had to say more about the good weather in the recovery area. "Yes, they told us that there's a beautiful moon out there."

"Yes," said Borman. "I was just [thinking] there's a beautiful earth out there."

In two hours they would give their last lunar press conference. Of the three astronauts, Borman had worried most about this moment. For weeks he had fretted about what they would say.

On that beautiful earth lived three billion people. Many had doubts about why his country had sent him here. Others wished that, instead of three Americans, two Russians had gotten there first. At least one third were about to celebrate one of the most holy religious holidays of the entire year.

Borman looked at the barren moon below him, and the distant earth beyond. For all he knew, he could be looking at a primeval universe immediately after its birth. The moon looked like a skull, bleached white by the hot sun. And though he knew there was life on earth, he could not see it.

And yet, on that distant orb people still lived, loved, fought, and survived. Looking at the three-quarters-full earth hanging in blackness, Borman sincerely wanted to tell everyone there what this journey had meant to him.

He hoped that he and his crewmates had found the way to do it.

## CHAPTER TEN

# "WHY DON'T YOU BEGIN AT THE BEGINNING?"

## C-PRIME

IN BERLIN THE WALL STILL STOOD. On its seventh anniversary, August 13, 1968, few demonstrated against its presence. Four thousand East German guards patrolled that grim border, and in the seven years since its construction over a hundred people were known to have been killed trying to burrow under or climb over it. During that time thousands more had fled successfully, including more than five hundred East German guards.[1]

Holiday passes between East and West Berlin had ceased. Except for emergency hardship passes, issued in the event of a death in the family, no West Berliners had been permitted to visit their East Berlin relatives in more than two years, since Easter 1966.[2] The East German government marked the wall's anniversary by praising its construction, noting how the outgoing tide of skilled and educated workers had stopped, and how this had benefited the East German economy.[3]

The week before East Germany celebrated the seventh anniversary of the Berlin Wall, George Low, manager of the Apollo program, returned from a two-week vacation in the Caribbean. Much had happened in the space program since he had first considered sending an Apollo spacecraft to the moon four months earlier.

The original schedule for the Apollo program required that six different tasks be achieved successfully before a lunar landing was attempted.

"A" had been the unmanned test flights of the Saturn 5, already accomplished with mixed results.

"B" had been an unmanned orbital mission of the lunar module, accomplished successfully in January 1968.

"C" would be the first manned mission of the Apollo spacecraft in low earth orbit. Apollo 7 was scheduled for October, 1968.

"D" would be the first manned mission in low earth orbit of the lunar module. This was Apollo 8, scheduled for early December with Frank Borman, Mike Collins, and Bill Anders.

"E" would repeat Mission D, but take the command module and lunar module to high earth orbit, about 4,000 miles. Apollo 9 was scheduled to do this in the early spring.

"F" would take the command and lunar modules into lunar orbit and test them. Apollo 10 would accomplish this task.

"G" was given to the actual lunar landing.

The sequence was carefully designed to test each component in the safest possible manner. No flight beyond low earth orbit would take place until the lunar module was ready. This way, if either craft failed, the astronauts would still have another vehicle to act as their "lifeboat" during the long journey back to earth. (When Apollo 13 failed in 1970, this was exactly what happened. Jim Lovell, Fred Haise, and Jack Swigert would have died in less than two hours had they not had the lunar module as backup.)

Unfortunately, since April problems had developed with this schedule. Borman, Lovell (replacing Collins), and Anders were ready for their December flight. In fact, after years of sitting on the sidelines, Bill Anders was primed and ready to go as the world's first lunar module pilot. In addition to spending thousands of hours in lunar module simulators, he and Neil

Armstrong had begun flying an ungainly, spiderlike aircraft called the Lunar Landing Research Vehicle (L.L.R.V.). Called a "flying bedstead" by some, the L.L.R.V. could only be kept aloft by firing a number of small thrusters at its base. In the earth's heavier gravity and windy atmosphere, the L.L.R.V. was a difficult craft to steer. In the six months prior to 1968 two had crashed, with the pilots ejecting safely.

Bill, however, was always able to land the thing, and by August 1968 he considered himself ready. Unfortunately, as ready as Bill Anders was, the actual lunar module was not. Construction was behind schedule, and the earliest the man-rated module for Apollo 8 would be expected to fly was February 1969.

During his vacation George Low had soaked up the sun on the beach and thought about this problem. For the last seven years Low had led the charge to send a man to the moon. In 1961 he had headed the committee that first proposed making the central purpose behind the Apollo program a lunar landing.[4]

He knew that time was of the essence. If NASA was going to meet Kennedy's deadline, they had to make every scheduled flight count. And there were other considerations. C.I.A. reports indicated that the Soviets were about ready to return to manned space flight, and that they could very well send men around the moon by year's end. In fact, the last unmanned test flight of the redesigned Soyuz spacecraft, named Cosmos 238, was set to launch before the end of the month.

Low knew that to send Apollo 8 up without a lunar module and merely repeat the Apollo 7 mission wouldn't accomplish anything. Nor did it make any sense to simply wait another two months for the lunar module (or LM, pronounced "lem") to be ready. Either case would probably delay a landing on the moon until 1970.

Low got an idea: Why not change Apollo 8's mission? Rather than wait for the unready LM, he would instead propose that Apollo 8, scheduled to blast off in December, would go to the moon.

When he returned from vacation on August 5th he immediately began canvassing people throughout the agency to see if his idea had any merit.

Wernher von Braun, builder of the Saturn 5, supported the proposal. As he noted, "Once you decide to man [a Saturn 5] it doesn't matter how far you go."[5] Since April he and his engineers had worked day and night trying

George Low in mission control.

to eliminate the problems experienced during the Apollo 6 flight, and now he felt they had.

The oscillations had been solved by adding "shock absorbers" to the rocket.[6] Discovering the cause of the rocket engine failures was more difficult, requiring thirty days of detective work. What they had found was that on the ground the engine's hydrogen fuel lines were strengthened by a thin frost layer of liquefied atmosphere, frozen to the surface of the pipe by the supercold fuel. In the vacuum of space, however, there was no atmosphere, and the unprotected fuel line would shake itself apart. Braun's engineers redesigned the lines with this in mind, and the problem was eliminated.[7]

The software people at mission control said that though they only had four months to write the programming for a lunar flight, they could do it.

The safety people saw no reason not to go to the moon. This was what the Apollo spacecraft had been designed to do in the first place.

When Deke Slayton asked Borman if he would be willing to take the Apollo 8 command module to the moon, he answered "Yes" instantly. Though

Borman had been willing to repeat his Gemini 7 experience for the good of the program, he really did not want to spend another ten days cooped up in a space capsule, even if its interior space had grown from that of a compact car to that of a mini-van. Going to the moon seemed infinitely more interesting.

Susan didn't tell Frank her concerns. NASA only had four months to plan this, the most ambitious space flight in history. They would be putting men on the Saturn 5 for the first time. They would be leaving earth orbit for the first time. And they would be flying the Apollo capsule for only the second time.

Bill Anders's only regret was that he would not be able to test fly the lunar module, for which he had been training for the last eighteen months.[8] He was now a lunar module pilot without a lunar module.

He told Valerie about the mission change, explaining that if he went he probably only had a fifty percent chance of coming back alive. With five small children and not a lot of money, the risks to his family were intimidating. And life insurance couldn't make up for the loss of a father.

They both recognized how significant an achievement a successful flight to the moon would be. And they both knew how difficult it would be for Bill to back out. He had to go, not merely because it was expected of him but because it was what he had been working for since the day he had joined the Air Force. Without her support his job would become much more difficult. Valerie disregarded her fears and backed him all the way. "Isn't this what you want to do?" she told him.

For her support Bill Anders is still immensely grateful. "She gave of herself for her husband, family, nation — with clear knowledge of the potential risks to *her*. She risked more and got less than I did. She's the hero."

Lovell, exhilarated by the news, immediately sketched out what was to become the mission's patch, a number eight (signifying both infinity and the mission number) with the earth in one loop and the moon in the other.[9] He came home and slyly told Marilyn that he wouldn't be able to go with her and the kids on an Acapulco Christmas vacation they had been planning for months. A hotel owner there routinely donated a suite to astronauts and their families following any flight into space.

Marilyn looked at him with annoyance. This was to have been their first family vacation in years. "Just where do you think you're going to be?"

"Oh, I don't know," he said sheepishly. "Maybe the moon."[10]

She was at first speechless. Then she saw the sparkle in Jim's eyes as he described what he and the other astronauts hoped to do, the long glide out from the earth, the arrival on Christmas Eve, and the plan to orbit the moon for twenty hours. It was Jim's childhood dream come to life. Like Susan and Valerie, she put aside her apprehension and doubts to stand with her husband.

Everyone was in line. All were in agreement that a mission to orbit the moon could be accomplished. Now they only had to convince James Webb, NASA's administrator, that the plan was feasible. When Thomas Paine, deputy administrator under Webb, broke the idea to him on August 15th, his reaction was one of shock and horror. "Are you out of your mind?"[11] he shouted. Sam Phillips, Director of the Apollo Program, who with Paine had passed on the news, noted that "if a person's shock could be transmitted over the telephone, I'd probably have been shot in the head."[12]

Though James Webb had been head of NASA for almost eight years, had shepherded the agency through its beginnings, and was certainly not afraid to send men to the moon, this mission was more risk than he wanted to take. Outvoted by everyone else in NASA, he decided it was time to let others run the show. Webb had already considered stepping down, and had even disucssed it with President Johnson. Both understood that for political reasons the space agency would probably be better served if a new administrator was in place before the next President took office in January.

On September 16th Webb met with President Johnson, and brought the subject of his resignation up again. This time he was shocked when Johnson immediately accepted, prompting Webb to make the announce-ment that very day.[13] On October 7th, just four days before the launch of the first manned Apollo mission, James E. Webb stepped down, and Thomas Paine took over. Webb went on to serve twelve years on the Board of Regents of the Smithsonian Institute, helping to guide this government institution as well as he had NASA.

Before his resignation, however, Webb allowed the planning for a possible Apollo 8 lunar mission to go forward, albeit in secret. On August 19th Borman and Kraft, along with a number of key flight planners, hammered out the tentative flight plan. Because of orbital mechanics and the need to arrive at the moon under certain lighting conditions, the launch would have to take place within a window of six days, beginning on Saturday December 21st. Once arriving in lunar orbit the astronauts would circle the moon ten times, then leave orbit on December 25th to return to earth on December 27th. They designated the flight as "C-Prime," since it only differed from the Apollo 7 "C" mission in that instead of circling the earth, Apollo 8 would circle the moon.

Later that same day, Sam Phillips announced that Apollo 8 might do more than fly in low earth orbit, that NASA was even considering a circumlunar mission. Besides this vague statement he said nothing. He knew, as did everyone else in NASA, that the Apollo 7 orbital mission in October had to be perfect before they could commit Apollo 8 to the moon.

REVOLUTION

The next day, August 20th, Soviet tanks rolled into Czechoslovakia. As they had in 1953 in East Germany, and in 1956 in Hungary, the Soviet Union had decided the policies of one of its neighboring states were unacceptable.

Eight months earlier Alexander Dubcek had become the leader of the Communist Party of Czechoslovakia, and had immediately moved to establish freedom of speech and to open his country's borders to the non-communist countries of Europe. Throughout that "Prague Spring," Czechoslovakian society bloomed. New independent parties formed. Large public demonstrations took place. By June a National Assembly felt confident enough to sanction these actions publicly. They declared censorship illegal, passed laws to protect the legal rights of the individual, and agreed to consider the right of opposition groups to form and petition.[14]

By August, the Soviets had had enough. Beginning just after midnight on August 20th, 650,000 troops and 6,000 tanks from East Germany, Poland, Hungary, and the Soviet Union poured across the borders. By 7:30 AM on August 21st Soviet troops were firing on demonstrators in the streets of Prague, and tanks had surrounded the headquarters of the Czechoslovakian Communist Party. Within hours Dubcek and six other leaders were arrested and taken as prisoners to the Soviet Union. In the first week over 20,000 refugees fled the country.

In 1968, however, it was Leonid Brezhnev ruling the Soviet Union and not Khrushchev. In the 1950's, Khrushchev had taken the seized leaders of the Hungary and East German revolts and had them shot. He then found men willing to run these countries the way *he* wanted, and enforced their rule with brute force.

Brezhnev, however, was more cautious. He did not shoot Dubcek and his followers. He did not use his army to make mass arrests. When the Czechoslovakians refused to accept the quisling government of his choice, Brezhnev simply reinstated Dubcek and the former Czechoslovak government and forcefully told them what they had to do if they wanted to stay in power.

They went along, albeit reluctantly. As per Brezhnev's orders, strict control over the press and media was re-instituted, and the Czechoslovakian foreign policy was once again closely coordinated with the other Eastern Bloc communist states. The withdrawal of Soviet troops, however, was left for later discussions.

Within weeks the Czech government announced the restoration of censorship and the forced disbanding of any non-communist political groups. Soon there were reports of writers and artists being arrested and beaten. By April 1969 Dubcek had resigned.[15]

The Prague Spring had ended.

In the West, however, protests and uprisings continued unabated. Dissent in 1968 had made this a very violent year. The riots following King's assassination in March and the take-over of the Columbia University campus in April and May had only been harbingers of later events.

In Paris and Rome, worker strikes practically shut down both cities. Hundreds were hurt in the French protests, with student protesters setting fires in the streets and fighting directly with police.[16] In the United States, students occupied buildings at Stony Brook, Boston, Oregon, San Francisco, and Northwestern Universities, to name just a few. A peaceful civil rights march in Washington in June turned violent. Storefronts were smashed and looted, rocks and bottles were flung at police, and for the second time in less than four months the National Guard was called in to patrol Washington, D.C. streets.[17]

On August 26th, one week after the invasion of Czechoslovakia, the Democratic Party opened its convention in Chicago to pick its nominee for the presidency.

The Republican Party had already chosen Richard Nixon as its candidate. Since his kitchen debates with Nikita Khrushchev, Nixon's career had run a roller coaster of failure and success. In 1960 he ran against John Kennedy for the presidency, and lost by a mere 118,574 votes.[18] In 1962 he ran for governor of California, and lost again. In his concession speech he had stood before the reporters and bitterly complained about how the press had always covered him. Then he told them that "You won't have Nixon to kick around any more, because, gentlemen, this is my last press conference."[19]

Now, in 1968, he was back in the running, having easily won the Republican Party nomination. Though he hadn't said what he would do about the never-ending war in Vietnam, other than to point out that he could handle the situation better than the Democrats, by September polls indicated that he held a solid lead for the presidency.[20]

With President Johnson's withdrawal, the race for the Democratic nomination had centered on three men. Robert Kennedy, who might have had the most support within the party, had been assassinated minutes after winning the California primary on June 4th. Eugene McCarthy, whose anti-war challenge to Johnson helped bring the president down, had won some primaries, but few within the party backed him.

Vice President Hubert Humphrey, a member of Johnson's administration, was actually in command. He had refused to run in any primary elections, knowing he could garner enough delegate votes to win merely by pulling party strings. He came to Chicago expecting to be nominated, and as expected, he was chosen as the Democratic Party's presidential candidate on the first ballot. In his acceptance speech he said nothing about what he would do to end the war in Vietnam, merely pledging vaguely that "the policies of tomorrow need not be limited by the policies of yesterday."[21]

The convention itself was the scene of more unruly demonstrations and violence. Like Columbia University, the hatred between protesters and law enforcement was palpable. Three thousand anti-war demonstrators attempted to storm the convention hall. Protest leader Tom Hayden told the crowd, "It may well be that the era of organized, peaceful and orderly demonstration is coming to an end and that other methods will be needed." In response, the police used tear gas, barricades, and batons in arresting almost six hundred protesters. About two hundred people were injured, including one hundred nineteen policemen.[22]

A little over two months later, on November 5th, Richard Nixon defeated Humphrey for the presidency, winning by the slimmest of margins, less than 500,000 votes out of over seventy million cast.[23] That same day, anti-war demonstrations involving several thousand people took place throughout the country. At the University of Michigan, five hundred students occupied campus administration offices in protest of the war. In New York eighty-four protesters were arrested when they marched through midtown Manhattan, blocking traffic.[24]

Many of the innumerable demonstrations and protests in 1968 were peaceful, reasoned, and acceptable under the American concept of freedom of speech. Many of the protesters disavowed the violence and intolerance of others. Many others choose to follow the principles of democracy rather than the force of tanks and guns, of bottles and bricks, and peacefully chose the ballot to change their country's leaders.

Nonetheless, the legacy of that time was one of intolerance. The Soviets refused to allow any free speech, and enforced their rule with an army of 650,000 men. In the United States, where free speech was permitted and democracy was the law, both dissenters and establishment too often chose force and violence as a means to impose their will. The police used any excuse to attack the protesters. "How would you like to stand around all night and be called names not even used in a brothel house," said one cop. Chicago Mayor Richard J. Daley called the protesters "a lawless, violent group of terrorists." The dissenters were equally offended when they couldn't get a majority of the country to agree with them. Their leadership, most of whom favored socialist or communist ideologies, repeatedly demanded the "spilling of blood" and for their opponents to be "pushed into the sea." Or as Tom Hayden shouted, "If blood flows, we must make sure it flows all over the city."[25]

Through this all, the Vietnam war raged on. By election day, almost 30,000 Americans had been killed in the southeast Asian jungles.[26] And no one could see an end to it all.

Overall, 1968 had not been a good year.

## APOLLO 8

To the contestants in the space race, however, all of these issues were distant tragedies. They had volunteered to do something noble, courageous, and bold, the last lap of the race had finally arrived, and neither side had time to think of anything else.

On September 15th the Soviets struck, proving that the C.I.A. reports had been correct. Zond 5 became the first vehicle to fly past the moon and be successfully recovered on earth, landing on September 21st in the Indian Ocean. A failure in the spacecraft's reentry guidance system caused it to miss Soviet territory. Despite being called "Zond," this was a Soyuz test spacecraft capable of carrying a human crew. This time, however, it merely carried a crew of turtles, flies, worms, and plants.

Three weeks after this Soviet triumph, NASA followed with Apollo 7, the first American manned mission in almost two years. Commanded by Wally Schirra, this eleven day orbital mission was the Apollo program's first shakedown flight.

Schirra, one of the original seven Mercury astronauts and the fifth American to fly in space, had not trusted the redesign work done by the investigation committee and Borman. Though scheduled as the commander of the first manned Apollo mission after the fire, he was blocked by Borman when he tried to take an active part in the command module's redesign. While Borman had sympathized with Schirra's concerns, he was also determined to keep the redesign focused. "They meant well," Borman wrote later of the numerous astronauts who wanted to take part in the redesign, "but their wish list was longer than a rich kid's letter to Santa Claus. If we had redesigned the spacecraft in accordance with everything they wanted, Neil Armstrong and Buzz Aldrin might not have landed on the moon for another five years."[27]

While most of the changes Schirra and the other astronauts desired were incorporated into the refitted spacecraft, he still chafed at not being able to supervise the changes personally. "We all spent a year wearing black arm bands for three very good men," he grumbled. "I'll be damned if anybody's going to spend the next year wearing one for me."[28]

Much like Gemini 7, Apollo 7 circled the earth again and again as the three astronauts tested the operation of the Apollo command module. Unfortunately, on the first day Schirra came down with a head cold, which he immediately passed to his fellow crew members, Walt Cunningham and Donn Eisele. Over the next week and a half, clogged sinuses combined with zero gravity and a pure oxygen atmosphere conspired to make this crew the grumpiest in history. At one point Schirra suddenly canceled a scheduled television broadcast by declaring, "The show is off! The television show is delayed without further discussion. We've not eaten. I've got a cold, and I refuse to foul up my time."[29]

Other than the colds and the ill-tempers that went with them, however, this flight had few technical problems. One of its most successful achievements was the use of the world's first handheld black-and-white video camera, built by RCA. Three times over the ten day mission the crew held live press conferences, joking and doing somersaults in zero gravity. It was here that the lighthearted Schirra of Gemini 6's harmonica and bells reappeared. He began the first telecast by holding up a card that read "Hello from the lovely Apollo room high atop everything." The third started with Schirra welcoming his audience to "The one and only original Apollo road show starring the greatest acrobats of outer space!"[30] Behind him Cunningham and Eisele did somersaults and pinwheels. The televised broadcasts were immensely popular, watched by millions, allowing the world to see what it was like for a person to float in space.

The investigation committee and all of NASA had been vindicated. Everything had worked so well that soon after landing NASA labeled this a "101 percent successful" mission.[31]

Four days later, on October 22nd, the Soviets joined the party. Georgi Beregovoi took off on Soyuz 2, the first manned Soviet mission since the tragic death of Komarov in April 1967. During Beregovoi's four day mission, ground control made extensive tests of their automatic earth-controlled docking system, maneuvering close to a Soyuz target vehicle but never successfully docking with it.

Then, on November 10th, five days after Richard Nixon beat Hubert Humphrey for the presidency, the Soviets launched Zond 6. Similar to Zond 5, this craft was sent on a course that would take it around the moon and then back to earth. Unlike Zond 5 and any previous Soviet mission, the Soviet press publicly announced Zond's mission before the flight was completed. Zond 6 was "to perfect the automatic functioning of a manned spacecraft that will be sent to the moon."[32]

With this announcement, NASA's decision became unavoidable. Everything had gone as planned on Apollo 7, and as Chris Kraft noted, everyone wanted to "beat the Russians' ass."[33]

On November 12th, even as Zond 6 was flying to the moon, Thomas Paine, the new NASA administrator, announced to the world that Apollo 8 would do the same. "After a careful and thorough examination of all the systems and risks involved, we have concluded that we are now ready to fly the most advanced mission for our Apollo 8 launch in December — the orbit around the moon."

As the news conference unfolded, the reporters couldn't help noting repeatedly the risks involved. One remarked how previous NASA planning had always included a lunar module for any flight to the moon, just in case the main engines failed. This one did not. Another wondered what the odds were for the mission to be successful. Several worried about the dangers of radiation and solar flares.

Sam Phillips squelched these doubts. "I'm not going to try and calculate a set of probability numbers or odds for you. I feel that we're ready for lunar orbit and that we have every reason to expect that we will be able to carry out the full mission and to succeed with it. If I wasn't convinced of that, I wouldn't have recommended such a mission in the first place."

Four days later, the three astronauts gathered at the Manned Spacecraft Center in Houston, Texas for a press conference of their own. Once again, many of the questions centered on the dangers and uniqueness of their particular mission: "Is this flight too risky after only one manned Apollo flight?" "What will be the most critical moment in the flight?" "How concerned are you about not being able to get out of lunar orbit?" "Have you considered the dangers of radiation and solar flares on your journey to the moon?"

At one point a reporter bluntly asked, "If you lose the main propulsion system, can you get out of lunar orbit and get back?"

"No," Bill Anders answered just as bluntly. "Once we are in lunar orbit, the main propulsion system has to operate."

Borman's response to all this was very matter-of-fact. "I was involved in some of the decisions that were made in re-engineering the Apollo [capsule], and I wouldn't get in the thing if I didn't think it was a safe vehicle." He also pointed out how absurd it was for some people to get "a little queasy" now about going when this had been the point of the whole program to begin with.

A number of reporters took a more philosophical approach to the mission, noting that the Apollo 8 craft would actually enter lunar orbit on Christmas Eve. One asked the three astronauts if they planned on "making a Christmas-type gesture from space."

Frank Borman tried to answer, but for once in his life he seemed to be at a loss for words. "Well, that's a secret, I think. We'll have to think . . ."

He paused. "We've already considered that and we'll have to . . ."

He paused again. "I think it would be inappropriate and . . ."

He paused a third time and then finally admitted, "Quite frankly, right now we don't have any idea what it might be."

In fact, only days before, Julian Scheer, NASA's assistant administrator for public affairs, had called Borman and told him that NASA was worried about what the astronauts would say when they gave their second press conference from lunar orbit. "We figure more people will be listening to your voice than that of any other man in history," he told Borman.[34]

It had suddenly dawned on everyone in NASA how important those words would be. For both the designers that created NASA and built the Saturn 5 as well as the military-trained test pilots that would fly her, this problem was far more challenging than building a rocket bigger than the Statue of Liberty.

On Christmas Eve 1968, three American astronauts were going to be orbiting the moon. What in God's name were they going to say?

## WORDS

On December 8th, the Apollo 8 astronauts moved into their crew housing at Cape Kennedy. These special isolated quarters, located on the third floor of an engineering and office building on the Cape, were provided by NASA in an attempt to reduce the chances that any astronaut would catch an ailment prior to launch.[35]

Each man had a small, private bedroom furnished much like a college dormitory, and all shared a single living room, a conference room, and a dining room. To compensate for the isolation, NASA provided the astronauts with a full-time chef. The man, who had been a tugboat cook, was considered "excellent" by Borman and "terrible" by Anders. Lovell merely grinned and ate the food.

As intended, they spent their days going over the flight plan, practicing maneuvers on the command module simulator, or reviewing the design of the rocket and spacecraft. For the first week they performed a dress rehearsal of the countdown, as well as practicing emergency procedures should the rocket fail during launch.

For exercise the astronauts mostly jogged across the building grounds. At night Borman would go outside and stare at the moon, wondering if they could actually hit their target from so far away.

Unexpectedly, however, the reporter's question on whether they were going to make "a Christmas-type gesture from space" had begun preying on Borman's mind. When Julian Scheer first enunciated the problem, Borman asked him if he had any recommendations about what the astronauts should say. Scheer's response was to the point. "I think it would be inappropriate for NASA and particularly for a public affairs person to be putting words in your mouth. NASA will not tell you what to say."

As commander, Borman found this problem much more complicated than flying a jet fighter or piloting a space capsule. Though it was his responsibility to find the appropriate Christmas Eve statement, he was not a poetic man, and didn't know where he could go to find the right words.

He considered using the prayer that Rod Rose had worked out for him, and rejected the idea. This was his own private message, for his own local church. It didn't speak for Lovell or Anders, or for the rest of the world. He needed something broader.

With only two weeks left before launch, Borman found himself thinking about the problem more than he would have liked.[36] Other more important worries, such as making sure his crew and his spacecraft were ready, were beginning to take a back seat to this seemingly minor challenge in public relations.

Borman thought that a message for world peace would be appropriate, but every idea he came up with seemed hollow. How could he, a military man, call for world peace when his nation was at that moment participating in a bloody war in Vietnam?

He asked the other two astronauts for suggestions. Anders, a practicing Catholic, proposed they make a Christian statement, possibly reading something about the meaning of Christmas. He suggested telling the traditional Christmas story.

Borman wasn't satisfied. More than a billion people would be watching their telecast, many of them not Christians. He needed a statement able to include them all.

Lovell was also at a loss. They talked about rewriting the words of "Twas the Night before Christmas" or "Jingle Bells," but dismissed these ideas instantly. "It was too flippant," Lovell remembers. "It did not match the event."

Borman called Susan and asked her if she had any suggestions. She also was baffled. "To please everybody, to send a message to the entire world at just that moment?" she said. "Wow, I can't imagine what that would be."

Getting desperate, Borman telephoned Simon Bourgin, whom he had met on his Asian tour after the Gemini 7 mission. Bourgin, a former newspaperman, worked for the United States Information Agency as science policy adviser. He helped plan and then participated in the tours, escorting the astronauts in their travels. Since 1966 the Bormans and Bourgins had become close personal friends. Frank found Si's gentle intelligence and refined artistic sense a refreshing and civilizing contrast from the world of military test pilots.

Bourgin agreed to think about the problem. To his mind, Apollo 8's trip "around the moon was not that big a thing. Other events would soon transcend it." Nonetheless, he knew that the flight, circling the moon on

Christmas, would have an "overwhelming importance because it would give us a kind of rebirth and spiritual renewal." He played around with these thoughts for a couple of days, but nothing he wrote seemed right. "[My words] didn't have the ring of epic truth — what [the astronauts] said was going to be as important as what they did."

Though he had been sworn to secrecy, Bourgin decided he needed help. One of his Washington friends, at the moment spokesman for the Bureau of the Budget, had once been a magazine and television writer. "I could talk about it with Joe because we had had a long friendship, and I trusted him totally."

Joe Laitin had been a war correspondent in Germany and the Pacific, and after the war had covered the Nuremburg trials for Reuters. Later he spent ten years in Hollywood as a freelance writer, doing articles for popular magazines like *Ladies Home Journal, Colliers Magazine,* and the *Saturday Evening Post.* Since 1963 he had been working the public relations beat in Washington.

Laitin's first reaction to the question was "Oh, this is a piece of cake, no problem."

In order to get some idea about how the astronauts would feel orbiting the moon, he asked Bourgin if NASA had any descriptions or pictures that could help him visualize what the astronauts would see. Since no one had yet gone to the moon, all Bourgin had was an artist's conception, which he immediately messengered to Laitin.

That night, after his wife and kids had gone to bed, Laitin sat down in his kitchen with his manual typewriter and propped the artist's conception up in front of him. The picture made the moon look like a gray cinder, and the earth was no larger than a tennis ball. Laitin imagined himself in lunar orbit, and suddenly felt a deep pang of loneliness for that tiny tennis ball. There lived "everything that was dear to me and that I loved." He started to type.

The first idea wasn't right, so he pulled the sheet from the typewriter and tried again. The second idea also wasn't right, so it too ended up as a crumpled sheet on the floor.

A few hours later hundreds of balls of paper littered the room. To his chagrin, Laitin was having the same problem as Borman: everything he wrote

Joe Laitin, teaching a world affairs course in Los Angeles, 1962.

sounded hollow in comparison with the war in Vietnam and the turmoil of the last year. Worse, Laitin recognized that because the astronauts were military men working for a government project, any platitudes uttered by them would seem like propaganda.

Like Bourgin, Borman, and everyone else, Laitin realized that the words somehow had to include the feelings and beliefs of as many people as possible. They also had to have a special ring, a majestic, almost biblical quality to them.

*Why not go to the Bible?,* he thought.

He went and got a Gideon Bible that he had once swiped from a hotel. Because the astronauts would be speaking on Christmas, *Christ's birthday,* he thought, Laitin immediately turned to the New Testament and began leafing through St. Mark's.

Still, nothing seemed right.

It was now four in the morning, and Laitin was getting desperate. At that moment his wife Christine came downstairs to see what was going on. Seeing her husband, a normally unreligious man, reading the Bible and surrounded by piles of crumbled paper, she immediately became fearful. "Joe, what have you *done*?"

He explained his problem. Christine Laitin had had as interesting a life as Joe Laitin. Born in Paris, she had been educated by the nuns at the Convent of the Sacred Heart off the Normandy coast. Rather than become a Roman Catholic nun, however, she joined the A.B.C. Ballet of Paris as a ballerina. During World War II she fought the Germans in the French Resistance. In 1947 she came to America as a war bride, and after two unsuccessful marriages, married Laitin in 1961. Their marriage would last until her death in 1995.

"If you want poetry you're looking in the wrong part of the Bible," she said. "You should look in the Old Testament for that."

"But it's going to be Christ's birthday," Laitin said, actually irritated.

"I don't care whose birthday it is, if you want that kind of language you have to look in the Old Testament."

Laitin was now really annoyed. He looked at the thick Bible in his hand. "It's four o'clock in the morning," he said almost angrily. "I wouldn't know where to begin."

His wife answered, "Why don't you begin at the beginning?" She remembered how the opening words of the Bible possessed that stark simplicity reminiscent of the wild Normandy coast where she had spent her childhood summers.

"You mean Genesis?" Laitin snapped as he sharply flipped the book to the first page.

Then he looked at the words. "My God, Christine, here it is."

Laitin sat down quickly and typed a memo to Bourgin. He pointed out that the astronauts should only read the verses if they seemed appropriate once in lunar orbit, and that the reading should be followed by silence, no trite comments from ground control. He added a short closing line for Borman to read, and sent Bourgin the memo. Then he completely forgot about it.

Bourgin immediately recognized that this was the right choice, especially in the context of the Cold War. "It gave the space race a certain moral and religious stature." He took Laitin's rough memo and retyped it as a letter. Because Borman, a man who wouldn't tolerate any distractions, hadn't wanted him to discuss the problem with anyone, he left Joe's name out. "Borman had enough to worry about at the time." After the mission Bourgin would tell him the whole story.

And like Laitin, Bourgin didn't think the whole issue was that important. He promptly forgot about it.

* * *

Borman looked at Bourgin's letter and instantly knew his problem was solved. Genesis was the perfect choice.

Lovell agreed. "You couldn't ask for better. The words were the foundation of most of the world's religions."

Anders by now had changed his mind. He had talked to Valerie about the problem, and had realized that whatever they said shouldn't merely celebrate the Christian holiday, but somehow be non-denominational as well. This seemed almost impossible, until Borman showed him Genesis.

Borman wasn't required to tell anyone what they planned on doing. As Lovell remembers, "It was before anyone thought of supervising these things." Borman did let a few NASA officials know about it, not to get approval but to make sure that the reading came off perfectly. Laitin's suggestion that no one say anything trite at the end of the reading required a little coordination. All agreed that once the astronauts finished reading, the ground communications should make no casual remarks like "Boy was that great!" Instead, there would be a long pause, and then a very respectful comment from public affairs officer Paul Haney.

Borman took Bourgin's letter, photocopied it onto fireproof paper, and inserted it into the flight manual. When Susan asked him what they had decided to do, he refused to tell her. All he would say was that he thought the choice was perfect.

## CHAPTER ELEVEN

# PILGRIMS
# TO THE MOON

*Being thus arrived in a good harbor and brought safe to land, they fell upon their knees and blessed the God of heaven, who had brought them over the vast and furious ocean, and delivered them from all the perils and miseries thereof, again to set their feet on the firm and stable earth, their proper element.*

—William Bradford, *1620,*
*describing his arrival at Plymouth Rock[1]*

EARLY EVENING IN HOUSTON. The sun has set. In the growing dusk the homes glittered with their bright Christmas decorations.

In front of one home a car pulled up and from it appeared a tall man dressed as Santa Claus. Jerry Hammack, a neighbor of the Lovells and in charge of the recovery operations at NASA, was making the rounds.

Many of the people who worked with him were Defense Department employees. Because the flight forced them to be away from their families during Christmas, he had decided he needed to do something to boost morale. He bought some presents, put on a Santa Claus costume, hoisted a large sack over his shoulder, and drove from house to house in an effort to bring good cheer to his co-workers and friends.

His last stop was the Lovell home. He especially wanted to cheer up Marilyn and her children. He rang the bell, and when Marilyn answered he entered with a "Ho-Ho-Ho." At first, two-year-old Jeffrey didn't know what to make of this big man with a white beard. He cringed in fear, so his mother held him in her arms so that Santa could give him a small gift. Then Jerry carried his sack into the living room, past the many friends gathered there (including his wife Adeline) and deposited boxes under the Christmas tree to be opened the next morning. He slipped out of sight for a moment to take off his costume, then joined the party. Soon the astronauts would be making their last telecast from lunar orbit, and no one wanted to miss it.

Similarly, in nearby El Lago, Valerie Anders, her children, and numerous friends were crowded around their two televisions. Ignoring the herd of reporters that surrounded her house, she and her friends had whiled away Christmas Eve day with idle chatter, listening to the astronauts as they circled the moon.

In Houston, Susan Borman, Fred, Ed, and Frank's parents were just now finishing up a quiet dinner with the Elkins. Afterward, both families went into the living room, where Susan discovered a surprise Christmas gift under the Elkins's tree. She tore the wrapping from the box and took out a dress that Frank had purchased for her just before leaving for Florida.

She held it up, baffled. The dress was at last six sizes too big. She began to laugh, and then found she had the giggles and couldn't stop. Soon everyone was laughing. Frank Borman, the man who instantly knew exactly what to do in any deadly emergency, couldn't buy a dress for his own wife.

\* \* \*

The astronauts meanwhile were hidden behind the moon on their ninth revolution. Anders and Lovell were now awake, and all three men hurried to get ready for what they knew would be the most watched moment of the entire mission. Despite all the discussion before launch about what to say, only now did they realize that they hadn't decided who would say what. After some hasty debate, Borman suggested that they first make a few comments each about their experience orbiting the moon. Then Anders would read the first four verses, Lovell the next four, and Borman would finish with the last four verses.

While they debated who should read what when, they also began packing up. From here on out, all that mattered was that one press conference and getting the S.P.S. engine ready for its last and most important use. "Because now she's going to take us home!" Borman exulted at one point. Away went most of the film cameras and all the paraphernalia that went with it. Also packed away was every loose item that might get flung about when the S.P.S. engine fired.

And they waited. Anders suggested that Lovell take out a map "so we can tell the folks what you're looking at." Then he asked Borman again what verses he should read, and took a look at the photocopy attached to the flight plan so that he knew where to find it.

With Anders guiding him, Borman oriented the spacecraft so they could see the lunar horizon. As the earth slowly rose above that razor sharp line, both Borman and Anders yelled, "Here it comes!" Anders snapped away with his camera, and within seconds they had reestablished contact with earth.

At mission control the room was packed. In one side booth sat Julian Scheer, NASA's assistant administrator for public affairs. Having refused to give Borman any suggestions for the telecast he had no idea what the astronauts were going to say. Now he waited for what he had calmly told Borman would probably be the most-watched television show in the history of mankind.

Si Bourgin and his wife were at that moment waiting in the airport bar in Houston. He had come to Houston for the launch, but was now heading home to celebrate Christmas Day with his family. He noticed that while a number of people were crowded around the television, many remained only slightly interested, eating and chatting among themselves as they watched.

Joe and Christine Laitin were at home. Laitin had never asked Si if Borman had accepted his idea, and he hadn't really thought about it much. He and his wife had finished dinner, and now were preparing for bed. He turned on the television so they could watch the lunar telecast before going to sleep.

Everywhere there was a television, people grew attentive, and the world stared as the camera on Apollo 8 finally turned on. Because the camera angle looked through the edge of the capsule's window, the initial image was distorted by spherical aberration. Below a tiny white blob, which was probably the earth, ran a number of horizontal streaks, one on top of another.

Borman spoke first. His crisp voice rang out clearly. "This is Apollo 8 coming to you live from the moon . . . We showed you first a view of earth as we've been watching it for the past sixteen hours." He paused. "Now we're switching so that we can show you the moon that we've been flying over at [seventy] miles altitude for the last sixteen hours."

As Anders shifted the camera, it accidently switched off, leaving earth with nothing but a series of gray bars. Without a TV monitor or eyepiece, however, the astronauts had no way of knowing this.

Borman continued, "Bill Anders, Jim Lovell and myself have spent the day before Christmas up here doing experiments, taking pictures, and firing our spacecraft engine to maneuver around. What we'll do now is follow the trail that we've been following all day and take you on through to a lunar sunset."

Ironically, the lack of a picture at this moment seemed to magnify their words. "The moon is a different thing to each one of us," Borman said. "I think that each one carried his own impression of what he's seen today. I know my impression is that it's a vast, lonely, forbidding-type existence, or expanse of nothing. It looks rather like clouds and clouds of pumice stone and it certainly would not appear to be a very inviting place to live or work. Jim, what have you thought most about?"

Jim Lovell's deeper voice hesitated slightly as he searched for the right words. "Well, Frank, my thoughts were very similar. The vast loneliness up here at the moon is awe-inspiring, and it makes you realize what you have back there on earth. The earth from here is a grand oasis in the big vastness of space."

"Bill, what do you think?" Borman asked.

Anders's voice, though similar to Frank's, had a softer, introspective quality to it. "I think the thing that impressed me the most was the lunar sunrises and sunsets. These in particular bring out the stark nature of the terrain."

About this moment the astronauts paused, and Jerry Carr radioed that they didn't have a picture. Lovell immediately spotted the problem and switched the camera back on.

As the image returned, it showed a bleak black-and-white lunar horizon line cutting across a rhombus-shaped window. To the viewers on earth, nothing could be seen but the lunar surface slowly drifting past, new craters creeping up over the horizon on the right as older craters disappeared off the window's edge on the left.

Anders continued. "The horizon here is very, very stark, the sky is pitch black and the earth, or the moon rather, excuse me, is quite light, and the contrast between the sky and the moon is a vivid dark line."

"Actually," Lovell put in, "I think the best way to describe this area is a vastness of black and white — absolutely no color."

Anders added, "The sky up here is also a rather forbidding, foreboding expanse of blackness, with no stars visible when we're flying over the ear—"

Note the rille in the center of the picture.

He caught himself again. "Over the moon in daylight. You can see that the moon has been bombarded through the eons with numerous meteorites. Every square inch is pockmarked."

Slowly the spacecraft drifted over the dead landscape. As he had earlier in the day, Anders did most of the talking, speaking softly and pausing often. Often he tried to describe the geological nature of what he saw. "There's an interesting rille directly in front of the spacecraft now, running along the edge of a small mountain, rather sinuous shape, with right-angle turns." A rille was a lunar surface feature, reminiscent of earth river valleys but deeper and

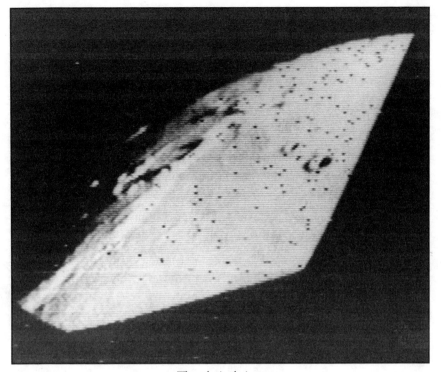

The televised view.

much narrower, with vertical walls and no tributaries. Scientists believe that they were formed from flowing lava.

Borman added a few more of his own thoughts. "I hope that all of you back on earth can see what we mean when we say it's a rather foreboding horizon. A rather stark and unappetizing looking place."

In mission control Julian Scheer noticed how, even as the astronauts spoke, the activity around him did not cease. "The room had a kind of hum," he remembered years later. Men still hunched over their consoles, studying the spacecraft's telemetry to make sure everything was working. Others moved through the room, passing information to and from each other. And all worked quietly and incessantly to keep this delicate spacecraft and its three passengers alive for just a few more days.

Now the spacecraft was moving toward lunar sunset, and the shadows were getting longer and more pronounced. Bill Anders opened the flight plan to the proper place. He glanced at the photocopy of Bourgin's letter. He began their closing comments. "We are now approaching lunar sunset and for all the people back on earth, the crew of Apollo 8 has a message that we would like to send to you."

Anders paused. Out the window the lifeless lunar surface drifted past. Nothing else moved. A long straight rille ran north-to-south along the sunset line. On one side was the bright harshness of lunar day; on the other was the utter darkness of lunar night.

At mission control the normal background hum of activity ceased. All movement stopped. All attention was on the TV and on Bill Anders's voice.

Anders began reading the first verse softly. "In the beginning, God created the heaven and the earth. And the earth was without form, and void; and darkness was upon the face of the deep."

In the Anderses' home complete silence fell. Valerie felt a thrill: *Bill was reading the Bible from the moon!*

"And the spirit of God moved upon the face of the waters. And God said, 'Let there be light.' And there was light."

At the airport bar in Houston, Si Bourgin noticed how the room had suddenly become very quiet. Even the bartender had stopped his work. All eyes were on that grainy black-and-white image on the television screen.

Bill Anders took a breath and read his last line. "And God saw the light, that it was good. And God divided the light from the darkness."

In Washington, Joe and Christine Laitin lay in bed, awestruck. Joe looked at his wife in disbelief. "That's the script I wrote!"

Christine looked back at him with a laugh, "You wrote?"

Now Bill Anders passed the flight plan to Jim Lovell, who began reading slowly. "And God called the light Day, and the darkness he called Night. And the evening and the morning were the first day."

In the Lovell household no one moved. Marilyn had had no idea the men would do something like this, and she was humbled. *They* must *be in God's hands*, she thought.

Lovell continued. His deep voice contrasted sharply with Anders's Western twang, and he struggled to make clear the strange wording of seventeenth-century English poetry. "And God said, 'Let there be a firmament in the midst of the waters, and let it divide the waters from the waters.' And God made the firmament and divided the waters which were under the firmament from the waters which were above the firmament. And it was so."

The shadows on the lunar surface had moved across the center of the image. Dark streaks cut across the white, into which nothing could be seen. With deep feeling, Lovell continued. "And God called the firmament Heaven. And the evening and the morning were the second day."

He passed the flight plan to Frank Borman, whose terse voice resonated keenly. "And God said, 'Let the waters under the heaven be gathered together unto one place, and let the dry land appear.' And it was so." Borman paused, glancing for a second at the steadily approaching lunar night.

At the Elkins house Susan Borman began to cry. *What a thing to do,* she thought. *How perfect a choice.*

Borman continued, "And God called the dry land *earth;* and the gathering together of the waters called he seas; and God saw that it was good."

Borman paused again. Outside the capsule window the shadows had become very long and wide. Lunar sunset was mere seconds away. He took a final breath. "And from the crew of Apollo 8," he said with utter conviction, "We close with good night, good luck, a Merry Christmas, and God bless all of you — all of you on the *good* earth."

At that moment Anders cut the television signal to earth, and television screens around the world suddenly turned to static.

In Washington, the phone at the Laitins' suddenly rang. It was Si Bourgin, calling from an airport pay phone. "Wasn't that great?" he said to Joe excitedly.

\* \* \*

Not long afterward, Marilyn Lovell and her children went for a nighttime walk, strolling through the darkness to admire the beautiful Christmas decorations that graced the homes of their neighbors. That year all the families of Timber Cove had decided to line their lawns with luminaries — small paper bags weighted down with sand and lit by bright candles.

While her children oohed and aahed at the pretty blinking lights, Marilyn could only think that at this moment these luminaries and Christmas lights had been blessed by the words of her husband and their father, spoken from a tiny capsule a quarter of a million miles from home.

These three men had stood on the fringes of human experience, tracing a warm line into the dark and cold emptiness of endless space, and had tried to bring more than mere life to that emptiness.

Their words were not original. They read words that had been written in the dim past by, as some believed, God Almighty.

Those words, however, expressed for these three men a heartfelt belief that the universe was more than mere energy and matter. Not only did a spirit lurk behind the veil of the terrifying black dark that surrounded them, it impelled them to live their lives a certain way, in a certain manner.

Their decision to read from the Bible also expressed, albeit indirectly, their passionate love of freedom. No one told them what to read, and in fact most of the officials at NASA and in the government were completely surprised by their message. And that was how it should be. Borman, Lovell, and Anders were free men, expressing their beliefs freely. While their government might have financed the journey, it could not tell them what to think or say once they got there. If these free men wished to pray aloud to the world's population as they circled the moon, so be it.

Their words also expressed their deep humility and abiding good will. They had been given this glorious opportunity to brag, and instead chose to pray, finding words that would include as many people as possible in the message.

Their voices, beamed across hundreds of thousands of miles by technology inconceivable ten years earlier, resonated with their country's roots as well. The Pilgrims had not merely gone to explore a new land — they had emigrated as families in order to build in that new land a human society.[2]

And so, like the Pilgrims, wherever the three men in Apollo 8 had gone they had brought their families, their religion, and their way of life. In

Houston they had found empty fields and built a community. Their lives had echoed the words of John Winthrop, leader of the first Puritan expedition, who as he and his fellow settlers first approached Massachusetts Bay in 1630 had urged them

> to do justly, to love mercy, to walk humbly with our God. For this end, we must be knit together in this work as one man. We must entertain each other in brotherly affection . . . We must uphold a familiar commerce together in all meekness, gentleness, patience, and liberality. We must delight in each other, make others' conditions our own, rejoice together, mourn together, labor, and suffer together, always having before our eyes our commission and community in the work, our community as members of the same body.
>
> [If we do this,] the Lord will be our God and delight to dwell among us . . . . He shall make us a praise and glory, that men shall say of succeeding plantations . . . we shall be as a city upon a hill, the eyes of all people are upon us.[3]

The community the three men in Apollo 8 wished to bring into the empty reaches of space was an American one, filled with a belief that given two strong arms, a willing heart, and the freedom to follow one's dreams, anything was possible. They, like the Puritans, had put their lives on the line to express this ideal.

Now the men orbited another world, farther away from home than any human had ever been, surrounded by airless space with only a week's worth of oxygen in their tanks. To put the final exclamation point on their powerful message, they had to get home, to their waiting families. And everyone knew that was not as easy a task as the astronauts and the engineers had so far made it seem.

## CHAPTER TWELVE

# "AMERICAN CHEESE"

WILLIAM ANDERS STARED OUT HIS WINDOW. "Look at that, look at that, Frank. Look at the earth!"

Frank Borman looked. "Yes," he said softly. This time they were watching the earth slowly *set* behind the moon. After a few seconds the blue-white globe slipped behind that sharp gray lunar horizon, and was gone. For the eleventh time, the astronauts were utterly alone.

It was now 11:40 PM. In Houston communications went silent, and a hush settled throughout the control room. Once again there was little anyone could do but wait. In thirty minutes, while still out of touch with the earth, the astronauts would fire the S.P.S. engine one more time for two minutes and eighteen seconds. If there was such a thing as a Santa Claus, no one would know until 12:34 AM on Christmas morning. If that S.P.S. engine performed as intended, Apollo 8 would then reappear from behind the moon, on course for home. If it failed, the men would be left trapped in orbit around the moon. The three astronauts would be condemned to die, unable to return and with enough oxygen for only a few short days.

To Susan Borman, T.E.I. was the worst moment of the mission. After dinner with the Elkinses, she and the boys had driven back to her home in El Lago. For an important event like T.E.I. she was obliged as the wife of the mission's commander to be home and accessible.

Valerie Anders joined her there. Earlier that week, the two women had decided that they would wait together at the Borman house during T.E.I. Valerie, exuberant and as fearless as the astronauts, had seen how worn Susan looked, and had offered to be with her at this crucial moment.

Together they now waited in Susan's kitchen, the house around them filled with friends and relatives, the squawk box hissing its dry chant of jargon. Susan sat with her eyes closed and her fists clenched. She still couldn't believe that all the engineering would work, that the S.P.S. engine would actually function perfectly on this first mission to the moon.

Valerie waited coolly, refusing to think negative thoughts. Instead she anticipated the moment the astronauts would reappear from behind the moon, on their way back to earth.

At the Lovell home, Marilyn also waited. After her walk with her children, she had come home to put her youngest ones to bed and join her relatives, friends, and neighbors in the family room. Like Valerie, she pushed the negative thoughts from her mind. All she thought about was that soon she would hear the voice of her husband, on his way home.

* * *

On the spacecraft, the astronauts waited for the engine to fire in quiet suspense. As soon as Borman finished the Christmas Eve press conference they began the long preparation for this last S.P.S. burn. For two and a half hours they worked. Sometimes they listened as Ken Mattingly dictated long sequences of numbers and data for them to enter into the on-board computer. Sometimes they scrambled about the cabin, trying to figure out where to cram the last pieces of equipment and garbage. Sometimes Borman and Anders went through another long checklist, making sure that everything on that complex instrument panel was set correctly. Sometimes Lovell floated in front of his sextant and tried to get navigational data, once again singing aloud as he worked.

And sometimes, they just joked and chatted about the events of the last few days. Once Borman speculated to Anders that it was the sleeping pill that made him sick on Saturday. Anders suggested that he might simply have had the flu, or an upset stomach. "Yes, I know," Borman answered, "But I never had it before, not even in the zero-g airplane."

Once Lovell wondered if the moon was made out of the same material as the earth. When Anders speculated that it probably was, both Borman and Lovell immediately needled him about his geological training. "Anders is going to tell you when he alights from the first LM," Borman kidded. "Pick up all the gold!"

"He *knows* it all, right now," Lovell added. "Didn't you hear him? 'Ye geologists of the world: I see a few grabens, a few slipped disks.'"

Once all three compared the real experience with the simulators in Houston. They were surprised when each admitted that the last week seemed more like being in the simulators than actually being in space.

As they slipped behind the moon, Anders took another picture of the earth. "We've got an earthset picture for *Life* magazine," he proclaimed as he snapped the picture.

And they each noted how amazing the last four days had been. "It's been a pretty fantastic week, hasn't it?" Borman said in wonder.

As they did these last chores, the clock moved past midnight and into Christmas Day. In mission control men stood silent with arms folded, waiting nervously. George Low sat in mission control in the V.I.P. lounge. This was by far his most fearful moment during the whole race to the moon. He hadn't been able to hide his worry in his conversation with Valerie Anders on Monday, and everyone knew that it was with his encouragement that NASA had hurried the program to send Apollo 8 moonward. As confident as he had been that everything would work, he was about to find out whether he had been right, or whether his decision had caused the death of three brave men.

Once again mission control sat in silence. Men whose shifts had ended hours earlier milled about near their consoles, unable to go home. Once again the man at capcom, this time astronaut Ken Mattingly, intoned a prayer-like litany into the microphone.

"Apollo 8, Houston . . ." An eighteen second pause.

"Apollo 8, Houston . . ." A twenty-eight second pause.

"Apollo 8, Houston . . ." A fifty-two second pause.

Now any delay in their reacquisition of the astronaut's radio signal would mean something was wrong.

In the Borman home, Susan Borman and Valerie Anders waited silently. At Marilyn Lovell's house, the room was hushed. The squawk box was silent. And then, as he had twenty hours earlier, Jim Lovell announced from the darkness that all was fine. "Houston, Apollo 8. Please be informed, there is a Santa Claus."

Amid cheers and a wave of almost audible relief, Mattingly sighed and responded, "That's affirmative. You are the best ones to know."

* * *

Christmas Day 1968 held a very special meaning for many people. In the Lovell home, Wednesday started with four young children swarming under the Christmas tree, tearing open the gifts that Santa had left them the day before. Twelve-year-old Jay was thrilled to get the race car set he had been wanting. Two-year-old Jeffrey was especially pleased with his toy helicopter.

As Marilyn Lovell watched her kids open their presents, her NASA press liaison came up to say that there was someone at the front door for her. Unsure if this was just another newsman wanting some meaningless comment, Marilyn opened the door slowly. Outside stood a man from Neiman-Marcus dressed in a chauffeur's uniform. He handed her a large box, and calmly climbed back into his Rolls Royce "delivery truck" and drove away. Attached to the box was a card that said "Merry Christmas and love from the Man in the Moon." The box was wrapped with royal blue foil paper on which a tiny model spacecraft orbited a styrofoam moon. Inside was a mink jacket. One last gift had made a late arrival.

When Marilyn went to church at 10:30, she wore the mink coat over her Sunday best, telling reporters that this was "her happiest Christmas ever."[1] When asked where the mink coat had come from, she said with a twinkle that it "came from the Man on the Moon, whoever that is. That's what it said on the package."

Leaving the moon behind.

In church Father Raish stood up before his congregation of astronaut wives and NASA employees and said, "Oh, eternal God, we commend to thy almighty protection thy servants James, Frank, and William, for whose preservation in space our prayers are offered."[2]

Valerie Anders had already received her Christmas present, the brand new color television set that she had been watching the mission on for the last five days. At 11:30 she took her five children to the Catholic church on Ellington Air Force Base where Father Vermillion led the congregation in prayer for the astronauts.

After church, Valerie took her kids to mission control so that she could send Bill her Christmas wishes. He answered her with his own blessing, promising her that he'd be home shortly.

In Susan Borman's home it seemed that she finally could eat and sleep. At 7 AM she went to church where she couldn't help expressing her relief to reporters. "This is truly a blessed Christmas," she said. Other than the oversized dress she had opened at the Elkinses', her Christmas presents remained wrapped under the Christmas tree, awaiting Frank's return.

Several hours later. The window frame is visible in lower right.

Not that these gifts mattered that much. To her, the only gift that counted was seeing her husband back on earth. She told reporters that "we'll be each other's big present."[3]

At St. Christopher's she played a tape of Frank reading Rod Rose's prayer, as well as of the three astronauts reading Genesis. Then Reverend Buckner, as Father Raish had done for Marilyn, delivered his own special prayer for the astronauts. "O eternal God, in whose dominion are all the planets, stars, and galaxies and all the reaches of time and space, from infinity to infinity, watch over and protect, we pray, the astronauts of our country."[4]

After services Susan went to mission control, taking her sons and Frank's parents. She didn't intend to pass any messages to Frank, knowing that would distract him. By now, however, things were going too well. When Mike Collins told Borman that he had a family of smiling Borman faces less than ten feet away, Frank Borman couldn't help grinning. He joked about how proud of him they'd be because he was exercising on the Exer-Genie, the device in the command module that simulated an exercise machine.

By now the astronauts had discovered that their Christmas hadn't been forgotten either. While weight considerations had limited each man's personal items to under seven ounces, they each found several small and very special gifts tucked away for them. Valerie had given Anders a tietack, a moonstone set in a gold number "8," while Marilyn Lovell had given Jim both a tie tack and cuff links, both set with moonstones.

Susan Borman had done something a little different. Just before Frank had left for Florida, the elderly mother of Margaret Elkins had given him a St. Christopher's boot tack worn by her deceased husband during World War I. This tiny medallion, about the size of a dime, had been pushed into the heel of his boot for good luck whenever he went into combat. The old woman had wanted Frank to have the same good luck her husband had had. Susan had secretly placed it in Frank's gift pack. While Frank might have felt distracted by any gifts from her, a heartfelt good luck charm from a faithful friend couldn't hurt.

Like Valerie and Susan, Marilyn made a quick trip to mission control so that she could send her Christmas greetings to Jim. She didn't stay long, however. Except for her kids and one or two close friends, her house was empty for the first time in days, and with the astronauts on their way home, she actually relished a quiet house in which to play with her children, listen to the squawk box, and anticipate the pending return of her husband. For her, the relief of their coming home seemed almost tangible .

As the day passed, communications between the ground and spacecraft became almost giddy at times. At one point astronaut Harrison "Jack" Schmidt read aloud a silly adaptation of "Twas the Night Before Christmas." "Frank Borman was nestled all snug in his bed, while visions of REFSMMATs danced in his head; and Jim Lovell, in his couch, and Anders, in the bay, were racking their brains over a computer display."

At another point Ken Mattingly, for no apparent reason, suddenly yelled "Eureka!" into the mike. Nor did anyone seem inclined to disagree with him.

All three astronauts slept. Jerry Carr gave the astronauts a very long and relaxed news report, much of which described how the astronauts' families had celebrated Christmas.

Borman and Lovell actually spent some time talking about football and whether the Houston Oilers could win their division the next year. The team had finished second behind Joe Namath's New York Jets, who in a few short weeks would be playing Johnny Unitas's heavily favored Baltimore Colts in the Super Bowl.

They did a mid-course correction, just to make absolutely sure they would hit the earth on target.

They did a television show. Held late in the afternoon on Christmas Day, this was a casual, hearty tour of the spacecraft, lasting over twenty minutes. Jim Lovell showed everyone how the Exer-Genie worked, Bill Anders ate a meal of cookies, orange juice, chowder and chicken, and Frank Borman described the computer and navigational system. As Borman later wrote, for this show they "hammed it up."[5]

After the telecast the astronauts discovered another surprise waiting for them inside the command module food locker. The Whirlpool Corporation, which held the contract for manufacturing the bland dehydrated meals that the astronauts had been eating for four days, had decided to produce a special meal for Christmas Day. The foil packages had "Merry Christmas" labels on them and were wrapped by the women in the Whirlpool mail room with green and red fireproof ribbons. Inside was real turkey meat and gravy. "It was the best meal of the trip," Borman wrote years later.[6]

Ironically, NASA had planned to use these more advanced food packages on Apollo 8. Borman had vetoed the idea, however, in his effort to eliminate as many risks as possible. He did not see any reason to try out new methods of food preservation when he and his crew were taking so many other first-time risks. Hence, for seven days the astronauts had endured the same unpleasant dehydrated food that Borman had eaten on Gemini 7.

Also part of this Christmas meal were three small bottles of Coronet V.S.Q. brandy, sealed in fireproof containers. Borman, still the unwavering no-nonsense commander, ordered the astronauts to put these back unopened. He was fearful that if anything should go wrong the liquor would be blamed.

And then, something did go wrong. After eating their Christmas dinner Borman decided to take a nap. He left Anders in the pilot's seat and slipped below the couches to rest. Lovell meanwhile slid over to his navigational station and began doing some more navigational sightings.

Jim, the navigational "concert pianist," started pressing buttons on the computer keyboard. He would sight a star in his telescope and tell the computer to align the spacecraft with it. The computer would then fire the thrusters to turn the capsule so that the chosen star was visible in Lovell's sextant. Lovell got so into this that he had the capsule swinging this way and that as he made star sighting after star sighting.

Suddenly Jim went "Whoa, whoa, whoa."

Mike Collins heard this and wondered what was going on. "Okay. Whoa, whoa. Standing by."

Lovell had accidently erased the navigational data from the computer. The inertial measuring unit (the I.M.U.), which Borman had insisted they leave on for the entire trip so that a manual realignment would not be necessary, no longer knew which way was up, and thought the spacecraft was instead back on launchpad 39A on Cape Kennedy. The computer, sensing that Apollo 8 was not oriented correctly, began firing thrusters. Anders, watching the systems, became alarmed as he saw the eight ball move more than he thought it should. The spacecraft was shifting drastically into a different position. He remembered Neil Armstrong's struggle to regain control of Gemini 8 when its thrusters had suddenly fired uncontrollably, and now wondered if this was happening to them. Anders fired a thruster to counteract.

The computer counteracted his action, and he counteracted the computer's counteraction. After a few seconds of battle Anders realized he didn't have a stuck thruster and let the computer stabilize the spacecraft to what it thought was a vertical position on the ground in Florida prior to launch.

Unfortunately, this orientation was useless for getting them home, since they no longer knew which way was up. This in turn made it impossible to align the capsule's heat shield properly when they reentered earth atmosphere, forty-one hours hence.

Not surprisingly, Frank Borman quickly awoke. While he and Anders sat and waited, Jim Lovell struggled to perform the manual realignment that Borman had hoped to avoid, resetting the I.M.U. so it knew exactly what attitude they were at that moment. Lovell looked out the window at the sun-washed sky and tried to identify a bright star in its constellation. Then he manually fired the thrusters to place this star in his sextant.

With a field of view of only 1.8 degrees, however, the sextant could not show him the entire constellation. Lovell could only guess that he had the right star in sight. After ten minutes, with Borman and Anders growing increasingly anxious, he managed finally to align Rigel and Sirius. After another fifteen minutes of tweaking, he was able to reset the computer so that it once again knew the craft's orientation in space.

Borman asked Collins if there was "any danger that this might have screwed up any other part of memory that would be involved with entry?" Collins told him no, but that the ground would keep checking.

Jim "Shaky" Lovell looked at his two partners with a sheepish grin and said, "Don't sweat it."[*]

Borman went back to sleep. Anders took over the controls again. Collins asked if Bill wanted him to pipe up music from the tapes Anders had provided the ground. "Go ahead," he said.

Suddenly he was listening to a choir singing "Joy to the World." He floated there for two minutes, captivated by the music. The choir began its second song, "O Holy Night." Anders was so mesmerized that he forgot to change antennas. As the spacecraft rotated in its "barbecue mode," he needed to periodically flip a switch to maintain communications with the ground.

As the active antenna rotated behind the capsule, the choir's voices began to distort and warble into incomprehensibility. Anders felt a prickly feeling at the back of his neck, not aware at first what was happening. It seemed to him as if everything were suddenly grinding to a halt, as if the powerful religious music of his world had no power over the vast universe he was now traversing.

Then he remembered the antenna and flipped the switch. The music came back clear and in its full glory. To Anders, however, he would never again hear that music without a prickly feeling at the back of his neck, and without wondering at the validity of the words.

---

[*] Ironically, Lovell was forced to repeat this unplanned manual emergency procedure once again during Apollo 13. On that flight, an explosion forced them to once again turn off the I.M.U. to save power, and Lovell and Fred Haise had to make a rough realignment using the sun and the earth. As Lovell notes today, "My training [on Apollo 8] came in handy!" See Lovell (1994), 283-284.

More time passed. As Ken Mattingly noted to Borman late that evening, "We're in a period of relaxed vigilance."

Borman responded, "We'll relax; you be vigilant."

Mattingly laughed. "That's a fair trade."

\* \* \*

On Thursday morning Jerry Carr opened the day with another daily news report, describing how, at the suggestion of Susan Borman, the families of the Apollo 8 astronauts had sent a prayer of thanksgiving to *Pueblo* Captain Lloyd Bucher and his wife. Carr also described how Bob Hope was once again entertaining the troops stationed in Vietnam, and how a so-called "gang" of high school teenagers in Ann Arbor, Michigan had gotten together secretly to cut through red tape and do good. Calling themselves the "Guerrillas for Good," Carr described how the youths had painted a bridge covered with obscenities, cleaned up trash along a river bed, and boarded up a condemned house.

In the afternoon, the astronauts gave their final in-space telecast, aiming the camera out the window to give the people of earth another view of themselves. It was obvious the home planet was slowly growing larger. The South American continent as well as Florida and the Caribbean could be seen under the swirls of clouds.

Watching this show from the control room was Marilyn Lovell and her two oldest children, Barbara and Jay. Barbara had been bothered the day before with what they all described as a twenty-four hour virus, though she felt well enough to come to mission control and see her father in space. In less than twenty-four hours the spacecraft would slam into the atmosphere, and Marilyn was beginning to feel increasingly tense once again as splashdown approached. It was the last hurdle she had to face.

Lovell stared at the approaching earth and couldn't help reflecting again on how tiny it seemed. "The earth looks pretty small right from here."

Bill Anders added his own thoughts. "As I look down on the earth here from so far out in space, I think I must have the feeling that the travelers in the old sailing ships used to have, going on a long voyage from home. And

The televised view of an approaching earth, December 26, 1968

now that we're headed back, I have the feeling of being proud of the trip but still happy to be going home."

Anders pondered his home world. To him, this was the most significant thing he had discovered on this journey to another planet. While the goal had been to explore the moon, he had found that the earth was by far more interesting. He was once again struck by its fragility, its smallness, and its jewel-like preciousness.

Valerie Anders was home during this show. Seeing that steadily increasing earth and hearing Bill talk about coming home made her "feel really good." While the astronauts were spending much of their day packing

A 70mm still picture of the same approaching earth, December 26, 1968.
The bright hazy glow in the center of the blue ocean is a reflection
of the sun off the ocean and atmosphere.

up for reentry, she had been doing the same, running errands about the house while listening to the squawk box.

Soon after, during a press briefing on her front lawn, the reporters asked her what she was doing with her day. Unable to think of anything interesting to tell them, she began to kid them with a silly story about how she had spent the *entire* day trying to set her hair. "The beauty shop is closed," she told them with a grin.[7] Ridiculous on its face, the story was even more absurd because her hair was naturally curly and she rarely had to do much with it. She thought they got the joke.

Unfortunately, one newsman didn't get it, and the next day newspapers across America described how Valerie Anders had spent her entire day setting her hair.

Susan Borman also spent most of Thursday listening to the squawk box and cleaning the house. Faye Stafford had offered to come over to help Susan before and during splashdown. Though the boys had volunteered to go pick her up, they had disappeared early that morning in Fred's car. Susan had no idea where they had gone, and eventually she sent someone else to get Faye.

When the boys finally turned up several hours later, Susan was startled to see fifteen-year-old Ed with a cast on his right thumb. "What happened to *you*?" she asked.

Ed shrugged. Late the night before, after his mother had gone to bed, he and Fred had gotten into a teenage fight in the bathroom. Ed had taken a swing at his older brother and hit him in the head. Instantly he felt something snap in his thumb.

Abruptly the fight ended, and the two boys became the best of friends. They couldn't let their mother know what had happened. For the rest of the night Ed sat in a chair in his bedroom, nursing his hand and keeping quiet. Just before dawn he and his brother got into Fred's car and drove to NASA where they had the hand X-rayed and a temporary cast put on.

Susan could only laugh. Knowing how distracted she was, they had simply taken care of the problem without her. *How blessed could a mother be?* she thought.

Once back from the doctor, the boys immediately changed into their hunting clothes, grabbed their shotguns, and climbed into Fred's car to go hunting again. Even with his right thumb in a cast Ed could still pull his gun's trigger, and they both wanted to get away from the rabble of reporters and visitors that engulfed their house.

Once again Fred gunned the engine as they raced out of El Lago. Once again a bunch of reporters followed in their own cars. Once again the boys drove through NASA to ditch them.

This time, however, the trick didn't work. Several journalists anticipated the boys and circled around NASA to pick them up as they came out
the back entrance. When Fred pulled into their friend's farm, so did a carload
of journalists.

One pointed at Ed's cast and suggested he fire a round for the camera.
Ed, at fifteen a crack shot and an expert hunter, obliged, bagging a field lark
at the same time. Then the two boys disappeared into the woods, leaving the
reporters behind in the front yard.

\* \* \*

Splashdown was scheduled for an hour before dawn, Pacific time, on Friday,
December 27th. This would be the first time NASA had attempted a landing
in the dark. When early flight planning had suggested keeping the astronauts
in lunar orbit several more hours so that the spacecraft could splash down in
daylight, Borman had fought this. "I didn't want to spend any more time in
lunar orbit than absolutely necessary. Any prolonging of the mission simply
increased the chances of something going wrong." When others argued that
a night landing meant no one would be able to see problems at splashdown,
he countered "What the hell does that matter? If [something] doesn't work,
we're all dead and it won't make any difference if nobody can see us."[8]

Borman's nonchalance, however, obscured the radical nature of Apollo 8's
return from the moon. Unlike every other space mission, they were not simply
slowing down from earth orbit. Instead, they were falling from an altitude of
almost 240,000 miles. At the moment they hit the earth's atmosphere, their speed
would be over 24,500 miles per hour — a world speed record.

In order to lessen their speed safely, NASA was going to use a concept
called the double skip trajectory. The craft was not aimed at the earth's dead center,
but at its atmospheric edge. Like a stone skipping over the water, the capsule
would plow through the upper atmosphere, leap up above it once, than plow back
down to fall towards the Pacific Ocean. If the angle of approach was too shallow,
however, the spacecraft would bounce out of the atmosphere and fly past the
earth, never to return. If the angle of approach was too steep, it would continue
to plow downward, burning up in a fiery conflagration.

This approach had only been tried four times before, once by NASA on Apollo 4 and three times by the Soviets with their Zond spacecraft. The Zonds had done it while returning from lunar orbit, but only once had the Soviets managed to make the concept work. NASA's last attempt at the double skip trajectory had been canceled during the unsuccessful Apollo 6 test flight.

Only minutes before hitting the atmosphere Borman blew the explosive bolts that held the service module to the command module. Though it had put them in lunar orbit and then sent them home, all three men were too busy to notice as it slowly drifted away.

Borman oriented the spacecraft so the rear heat shield faced downward. Out their windows now they could see only sky and the horizon line. And then, six minutes before hitting the atmosphere and just as predicted, the now-distant moon made one final short appearance in Bill Anders' window.

A few minutes later Borman handed the controls over to the computer. Only if things went wrong would he take over and fly the craft manually.

At the same time the astronauts could see a hazy glow building up outside their windows. The spacecraft had finally returned to earth and touched its life-giving atmosphere.

Now, however, that atmosphere was deadly. The heat of reentry would exceed 5,000 degrees Fahrenheit.

The spacecraft began to grind through the air. The deceleration caused the astronauts to feel gravity, and because they were flying tail first, it made them feel like they were lying in their couches upside down.

"Hang on!" cried Borman, as the spacecraft shuddered earthward.

"They're building up!" called Lovell.

"Call out the g's," Borman said. In the excitement of reentry, Lovell had forgotten that it was his job to announce the g forces as they increased.

"We're one g," he now said calmly. This was equivalent to the force of gravity on the surface of the earth.

And then suddenly the pressure increased. "Five!" Lovell shouted.

Twenty more seconds passed. "Six!" he yelled. The astronauts now weighed about a half ton each.

And then the pressure dropped as fast as it had started. The spacecraft was now skipping back out of the earth's atmosphere. For another minute they rose, the g forces dropped, and then they began falling once more. "Three," Lovell called out as the g forces built up again.

For five minutes the cone-shaped spacecraft roared through the earth's atmosphere, flying over more than 1,500 miles of earth terrain. Inside the capsule everything was lit by the soft intense light emitted by the glowing heat shield below them and the bow shock of ionized gas that surrounded them. As expected, their radio communications were now blacked out. To Borman it was being "in a neon tube."[9] Anders could see chunks of the melting heat shield fly past his window. He wondered if too much was breaking off, and if soon he would begin to feel heat against his back.

Three minutes later they dropped below 100,000 feet elevation, and the glow around the spacecraft dissipated enough for them to regain contact with mission control.

At 30,000 feet the computer blew the parachute cover off with a bang, followed immediately with another bang as it released the drogue chutes, small parachutes for stabilizing the craft and cutting its initial high speed in preparation for the main chutes. Seconds later an air vent opened and there was a loud blast of air as the cabin pressure equalized with the earth's atmosphere.

At 10,000 feet a third bang signaled the release of the main chutes, but in the darkness none of the astronauts could see whether this had actually happened. Like a pilot flying blind and totally dependent on his instruments, Borman noted that according to his display they were now dropping at about twenty-five miles an hour. "We're going down very slow," he told the rescue helicopter pilot.

Now came their last task. NASA had learned that the Apollo command module was somewhat top heavy in water, and would tend to float upside down. Furthermore, the parachutes tended to pull the craft sideways, helping to flip it.

In order to prevent this, Borman would cut the parachutes free immediately after splashdown, and then press a switch that would inflate three large balloons stored in the capsule's nose, keeping the craft right side up.

Unfortunately, the darkness made it impossible for the astronauts to judge when they would hit the water, and the impact was so strong that

Borman was staggered. On top of this, he was suddenly dunked by a surge of water, "from where we had no idea."[10] Consequently he released the chutes too late, and the spacecraft was pulled upside down.

As they hung there in their harnesses (with trash that they had stored under their seats raining down upon them), Borman hit the switch to inflate the balloons. After a few minutes the spacecraft flipped upright with a violent bounce.

Now they had to sit and wait for dawn. The Navy had already located them, but could not drop any swimmers into the water until daylight due to sharks. As they sat there Borman found himself quickly getting seasick. As he threw up, his crewmates, both Naval Academy graduates, couldn't resist making fun of the "West Point ground-pounder." The commander was no longer in charge, and his two crewmen took full advantage of his miserable condition to tell him about it. As Borman later admitted good-naturedly, his crew "performed admirably after we were on the water, [while] the commander was taking a vacation."

After about half an hour the sky had brightened enough for divers to hit the water and attach a flotation collar to the capsule. As everyone waited for the hatch to open, someone in a rescue helicopter radioed a question to these three first-time lunar explorers: "Hey, Apollo 8, is the moon made of green cheese?"

"No," Bill Anders said instantly. "It's made of *American* cheese."

## CHAPTER THIRTEEN

# THAT
# WAS THEN

## THE SQUARES

FRANK BORMAN STOOD ON A SMALL WOODEN PLATFORM overlooking a gray, eight-foot-high concrete wall. Beyond the wall he could see an open dirt strip filled with rolls of barbed wire and patrolled by machine gun-toting soldiers. Beyond them were gray abandoned buildings, their windows cemented shut.

Scattered along the near side of the wall were plaques. Each commemorated the place where a refugee had died trying to cross the death strip. By now there were over hundred such plaques. One was for Peter Fechter.

The date was February 11, 1969. Twenty years after he had flown into Berlin on a sack of coal, Frank Borman had finally returned to Europe. No longer an unknown cadet attending West Point, he now brought his wife and family with him. And he came as a famous American hero who had helped take the human race to the stars.

In the six weeks since splashdown a lot had happened to Borman, Lovell, and Anders. Within an hour of landing they had been airlifted by helicopter to the *U.S.S. Yorktown*, where they stepped onto the deck to the cheers of hundreds of Navy sailors. For Borman, the personal satisfaction and exultation reached its peak at this moment. "I wish I could describe the feeling of euphoria I felt," he said thirty years later. "It was the greatest feeling in the world."

On board ship they were wined and dined by the captain and crew. President Johnson congratulated them by phone. Then they were flown to Hawaii, where thousands of people flocked to see them. Such celebrations were to be expected after a successful space flight.

Then they landed at Ellington Air Force Base, just south of Houston. Because it was 2:30 in the morning, they only anticipated a small contingent of newsmen, NASA officials, and their families there to greet them.

Instead, the airfield was packed with thousands upon thousands of well-wishers. Many were co-workers at NASA, out to celebrate. Many others were mere strangers, exuberantly cheering the astronauts for what they had done.

Though it was the middle of the night, some neighbors had disassembled the American flag that they had built and set up on the Anderses' lawn in order to reassemble it at the airbase. They plugged it in, bathing the returning astronauts in its red, white, and blue lights.

This was nothing like Gemini 7, thought the Bormans and Lovells. The Anderses were simply astonished at the size of the crowd. The astronauts placed Hawaiian lei's around the necks of their wives, and embraced their families. Alan Anders, eleven, suddenly began crying.

Everywhere the astronauts went they were feted with honors and applause. In the month following splashdown the astronauts and their families went on whirlwind tours of New York, Washington, Miami, Chicago, and Houston, giving speeches and attending parades and parties.

In New York, Governor Rockefeller arranged for them to stay in the penthouse suite of the Waldorf-Astoria. To the Anderses' boys, Alan and Glen, this forty-two-story building seemed the tallest in the world. They amused themselves by flinging grapes out the window, seeing if they could hit pedestrians. "The best part was that we could drink all the soda we wanted," Alan remembered.

Soon the letters and telegrams began pouring in, numbering in the hundreds of thousands. The families didn't know how to answer them all.

One in particular struck Borman as especially poignant. "To the crew of Apollo 8. Thank you. You saved 1968."[1]

President Nixon, who had just taken office, asked Borman to go on a goodwill tour of Europe, and Anders to take over as Executive Secretary of the National Aeronautics and Space Council.

And then came the awards. The three men were given the Distinguished Service Medal from President Johnson, the Hubbard Medal from the National Geographic Society, the Collier Trophy from the National Aeronautic Association, and the Goddard Trophy from the National Space Club, to name just a few.[2]

*Time* magazine named the three astronauts its "Men of the Year." This honor, announced in the magazine's first issue of 1969, had been given each year since 1927, and was awarded to those individuals who had wielded the most influence on human history in the preceding year, "for better or for worse."[3] Prior to launch, the editors at *Time* thought they would name "The Dissenter" as 1968's "Man of the Year." The demonstrations, the violence, the discord, had dominated almost every day's headlines.

On December 28th, however, they had changed their minds, writing how the flight of Borman, Lovell, and Anders had "overshadowed—even if, in the long view of history, it did not cancel out—many of the most compelling events of the year." *Time's* editors noted that while the utopia the dissenters craved would never be found on the moon,

> ...the moon flight of Apollo 8 shows how that Utopian tomorrow could come about.
>
> For this is what Westernized man can do. He will not turn into a passive, contemplative being; he will not drop out and turn off; he will not seek stability and inner peace in the quest for nirvana. Western man is Faust, and if he knows anything at all, he knows how to challenge nature, how to dare against dangerous odds and even against reason. He knows how to reach for the moon.[4]

Not everyone celebrated the journey of Borman, Lovell, and Anders. Madalyn Murray O'Hair, the woman whose court suit had banned prayer in the public schools, immediately registered a complaint about the prayers the astronauts had read. "Christianity, you know, is a minor religion," she noted as she announced a letter campaign to ban public prayer in space.[5] Within months she had collected 28,000 signatures on a petition, and went to court demanding a ban on any Bible reading by any U.S. astronaut or any government employee while on duty.[6]

The court suit only helped fuel the wave of support for the mission. In the next year NASA received more than 2,500,000 letters and petitions objecting to her suit and supporting the right of American astronauts to exercise their religious beliefs publicly. Eleven months later the court finally agreed, dismissing her suit. "The First Amendment does not require the state to be hostile to religion, but only neutral," said one judge.[7]

In the Soviet Union the success of Apollo 8 also met with mixed feelings. The day before Borman, Lovell, and Anders reached lunar orbit, Georgi Petrov, head of the Institute of Space Research of the Soviet Academy of Sciences, wrote in *Pravda* that while he wished the astronauts well, he believed Soviet methods of space exploration were inherently safer.

> The Apollo 8 system is distinguished by the fact that the crew apparently plays the main role in controlling the craft. Soviet scientists and designers have been working on systems in which man's control of the spaceship is completely duplicated by automatic devices. . . . It seems to me that such a control system and preliminary testing of the entire flight program by automatic stations before sending off a manned spaceship insures greater safety.[8]

While Petrov's description of the Apollo 8 system was naïve, his honest description of the Zond and Soyuz automatic control systems was especially ironic. The Soviet decision to make their spacecraft a technological reflection of their society, in which the ground controller was the centralized authority dictating what the ordinary citizen cosmonaut could do, had helped guarantee that they would lose the race. The complexity of building such an automatic spacecraft made it impossible to get it ready in time.

Petrov also said that the Soviet Union's space goals were to establish permanent orbital stations. Unable to beat the United States to the moon, the Soviets had decided to make believe they had never intended to go. Instead, they now claimed that their efforts had been aimed towards a different objective.

This was a lie. Their aggressive effort to beat the Americans to the moon had failed not only because their spacecraft was unnecessarily complex, but because their increased caution after the death of Komarov had stymied them.[9]

The Soviets had insisted that no manned mission to the moon could take place until at least one unmanned robot mission was able to duplicate the flight, without problems. Thus, though both Zond missions in September and November 1968 had been able to send a spacecraft around the moon and return it intact to earth, both flights had failures that prevented them from meeting the mission criteria.

So, despite having scheduled a lunar mission for December 4th with a crew of cosmonauts trained and willing to fly it, Brezhnev's government refused to take the chance. Apollo 8 was therefore able to get to the moon first.

For the Soviet engineers and cosmonauts it was an agonizing experience watching the flight of Apollo 8. "It is a red letter day for all mankind, but for us it is marred by a sense of missed opportunities," wrote Nikolay Kamanin, director of cosmonaut training, in his diaries. "Americans are flying to the moon and we have nothing to counter their exploit. The most dismaying thing is that we cannot tell the truth to our people. We try to write and speak about the reasons for our setbacks, but all our attempts are mired in official bureaucracy."[10]

Less than a month after the flight of Apollo 8, the Soviet government canceled all plans to send any cosmonauts to the moon. The race to the moon was officially over. America had won.

* * *

In the United States the celebrations continued. All three families attended the Super Bowl in Miami. The Borman boys, high school football players, were given jobs as ball boys, and the three astronauts led the crowd in the

Pledge of Allegiance prior to kick-off. They then watched Joe Namath's A.F.L. New York Jets upset Johnny Unitas and the N.F.L.'s Baltimore Colts 16-7.

Bill Anders's picture of that breathtaking first earthrise was placed on a six-cent U.S. stamp. At first the government planned to issue the stamp with no text, but decided to add the words "In the beginning . . ." in response to the thousands of letters that poured into the Post Office.[11]

The three astronauts spoke before a joint session of Congress. Borman spoke last and longest, as was his right as commander of the mission. "Exploration is really the essence of the human spirit, and I hope we never forget that," he told a cheering room of legislators.[12]

Thomas O. Paine, acting administrator of NASA, couldn't help exulting at Apollo 8's success. "It's the triumph of the squares," he said. "The guys with computers and slide-rules who read the Bible on Christmas Eve."[13]

## CHANGES

And yet, did they triumph? In retrospect the great irony of Apollo 8 is that in the three following decades, society decided to draw from this space flight a completely different message, one that the astronauts and the other "squares" that sent them to the moon would not necessarily have endorsed.

Seven months after Apollo 8, Neil Armstrong and Buzz Aldrin landed on the moon, fulfilling Kennedy's pledge that "whatever mankind must undertake, free men must fully share." Before launch Buzz Aldrin had decided that, like Borman, Lovell, and Anders, he too wanted to give thanks to his God. He had worked out a short ceremony with his local church pastor, who had provided Aldrin with a tiny Communion kit with a silver chalice and wine vial "about the size of the tip of my little finger."[14]

But something had changed in the ensuing months. Though Julian Scheer once again told the astronauts they were entirely free to say whatever they wanted, at least one official at NASA actually advised Aldrin against saying his prayers in public.[15] As Aldrin noted many years later, some at the agency wanted to avoid any "adverse publicity from people like [Madalyn]

O'Hair."[16] Her complaints and court suit over the astronauts' reading of Genesis seemed to have intimidated the agency.

So, when Aldrin asked a breathless world several hours after the lunar landing "to pause for a moment and contemplate the events of the past few hours, and to give thanks in his or her own way," he was euphemistically telling the world that he was at that moment taking communion. He poured the wine and read silently words from the Book of John.

> I am the wine and you are the branches.
> Whoever remains in me and I in him will bear much fruit;
> For you can do nothing without me.

Yet no one, not even Aldrin's wife, knew that he was praying at this particular moment. Astonishingly, the objections of a few Americans had served to muzzle another American's freedom to speak.[17]

Even more surprising was how little anyone protested. No one in NASA, including Aldrin, saw anything wrong with an astronaut censoring himself because others disliked the public expression of prayer. While Borman, Anders, and Lovell had assumed they had the freedom to say whatever they wished, Aldrin did not. Something had changed in NASA in the ensuing six months.

In the ensuing decades the social pressure on astronauts to censor themselves only worsened. Today, when philosophical words are spoken in space, they rarely have the same sincere ring of honesty as the words of the Apollo 8 astronauts. Astronauts today all too often toe the party line, fearful that if they speak too boldly, their bosses on earth might ban them from further missions in space.

Instead, the politicians speak, using space as a platform to promote themselves. Prior to Apollo 8 no American President had ever spoken directly to an astronaut while still in orbit, as Khrushchev had done. Such an action seemed too self-serving and propagandistic: the astronauts had dangerous work to do, and it seemed unseemly for a mere politician to insert himself unnecessarily in that work.

By the Apollo 11 landing on the moon, however, that changed. Richard Nixon, who had done almost nothing to conceive, design, build, and fly the American space program, spent two minutes on the phone with Neil Armstrong and Buzz Aldrin as they stood on the moon, mouthing empty platitudes while these men's lives were at risk. Nor has Nixon been the only one to do this. Practically every President since Nixon has thought it acceptable to use space as his own personal bully pulpit.[18]

\* \* \*

Other changes were as fast and as immediate. In January, 1969, the *New Republic* published a short article by Ralph Lapp noting that no one knew exactly what would follow a lunar landing. "Slum-dwellers, awed by Apollo 8 and subsequent flights, may conclude that they should receive space age benefits. If the United States can accomplish such wonders, they may reason, can't we have decent housing, good schools, and a better life?"

Then in March Tom Wicker wrote on the *New York Times* op-ed page:

> The vision, skill, courage and intelligence that have gone into the space program ought to shame mankind—and Americans in particular. Because if men can do what the astronauts and their earthbound colleagues—human beings all—have done, why cannot we build the houses we need? Why must our cities be choked in traffic and the polluted air it produces? . . . Why does every effort to remove slums and rebuild cities bog down in red tape and red ink?[19]

Neither Wicker nor Lapp were asking that the space program be replaced by other government programs. Yet others took their words and came to that very conclusion. On May 20th, even as Apollo 10 and its three astronauts were halfway to the moon on the last dress rehearsal for the planned landing of Apollo 11 in July, Edward Kennedy stood up at the dedication of the $5.4 million Robert Goddard Library at Clark University to declare that Americans should slow their exploration of space, diverting the money instead to earth-based social programs.[20]

Nor was Kennedy alone in his demand. In the months following the Apollo 8 mission, the calls to reduce the American space program were incessant and many. For example, when the American Association for the Advancement of Science held a panel discussion on the space program at its December 1969 meeting, more than one hundred demonstrators also gathered to protest what they called a "moondoggle" and "our twisted national priorities."[21] The organizers of the protest later referred to NASA as a place "where the most outrageous forms of waste for profit are perpetuated . . . and used to divert attention from the obvious neglect of peoples' needs."[22]

Frank Borman himself got a personal up-front look at this groundswell of hostility. That spring President Nixon asked him to visit a number of colleges and universities to explain what the space program was about. In writing about that college tour years later, Borman bluntly described it as "a disaster." Astronaut Borman, a former test pilot and military man, found himself the target for the student anger at the continuing Vietnam War. Often students refused to let him speak, drowning him out with boos and catcalls. At Columbia, birthplace of violent student protest, the audience threw marshmallows at him, while others climbed onto the stage in gorilla costumes.

To Borman, one night at Cornell was particularly painful. "I wanted to talk about space, not an unpopular war, yet I ended up becoming almost an apologist for the military-industrial complex in the eyes of my radical-minded audiences who didn't want to hear about space."[23] Almost all the students Borman met, many of whom belonged to the S.D.S., seemed to reject the idea of exploration. "'How can you spend all this money going to the moon when there are so many poor, so many economic inequities, so much poverty?'" they asked Borman.[24] Rejecting the space program, the students, many of whom admired communism, instead wished to use the government to re-shape society, thereby solving other national problems that they considered more important. As former S.D.S. President Todd Gitlin wrote, "Much of the New Left . . . scorned the 1969 moon landing as a techno-irrelevancy if not an exercise in imperial distraction and space colonialism."[25]

Ironically, the United States had entered the space race to prove that in the competition between the totalitarian Soviet Union and the free and capitalist United States, individual responsibility and private enterprise could do it better.

"Ask not what your country can do for you, ask what *you* can do for your country," John Kennedy had said in his first inaugural address. According to this ideal, each citizen was individually responsible for doing what *he or she* could to make society better, *not* the government. Consider how the 1960's American space program operated: NASA was run so much from the bottom that the presence and influence of Washington is practically invisible. The hard scheduling and engineering decisions were instead worked out by the ordinary lower-echelon workers thinking independently in the field. Not surprising then that the last person to find out about the prospect of sending Apollo 8 to the moon was NASA's administrator, the man who was supposedly in charge at the top. James Webb might have been responsible for making the final and essentially political decision, but unlike his Soviet counterpart he interfered little with design and engineering problems.

NASA, however, was a *government program*, and its success helped prove — not only to the demonstrators that Borman faced, but to an entire generation — how it was possible to use government to solve society's ills. "As revolutionary visions faded, many became crisp professional lobbyists: environmentalist, feminist, antiwar," wrote Gitlin of the aftermath of the 1960's. "Most were willing to think of themselves as unabashed reformers, availing themselves of whatever room they found for lobbying, running for office, creating local, statewide and regional organizations."[26]

While Americans had often used local government to achieve their ends, and while the first half of the twentieth century had seen continuous growth in the use of the federal government to wield change, the 1960's saw a burst of federal activism that was possibly the largest in the country's history.[27] The success of the space program, though certainly not the sole cause, surely helped weaken resistance to centralizing American political power around the national government.

Nowhere was this process more obvious than in the environmental movement. As nature photographer Galen Rowell said in 1995, Bill Anders' photograph of an earthrise over a barren lunar surface was "the most influential environmental photograph ever taken."[28] Every edition of *The Whole Earth Catalog* displayed this picture on its inside cover, describing it as

The famous Apollo 8 picture of earthrise over the moon that established our planetary facthood and beauty and rareness (dry moon, barren space) and began to bend human consciousness.[29]

Even the S.D.S., which had never shown any interest in technological matters except to condemn companies like General Electric and the Rand Corporation for doing research for the Defense Department, became suddenly aware of environmental issues soon after Apollo.[30]

Nor were political movements alone in discovering a new awareness of our home world. As Carl Sagan wrote in 1975, "It is impossible to look at such pictures without acquiring a new perspective on our world."[31] Humanity had seen the earth for the first time as a *planet*, the globe's blue seas and swirls of white clouds giving it a colorful beauty and exuberance seen nowhere else. And as the only place in the universe known to sustain life, this "fragile Christmas tree ornament" beckoned to both the astronauts and to the population at large as a safe haven, a place that must be protected from harm at any cost.

Almost every astronaut to go to the moon after Apollo 8 said that the earth appeared delicate and fragile. Borman, Lovell, and Anders, however, said it first, and they said it when the greatest number of people was listening. Their words, and the images their took, shook society as much as Khrushchev's words and actions had in the 1950's.

Combined with an increased social desire to use government for moral ends, the images brought back from Apollo motivated environmental activism as never before. Though the movement had existed long before Apollo 8, environmental policy until this time had rarely used the federal government to enforce regulation. The National Park Service was created, and certain scenic areas of the United States were set aside for posterity, but national environmental legislation was uncommon and limited in scope.

Within two years of this first lunar voyage, however, the country celebrated its first Earth Day, and Congress passed both the National Environmental Policy Act and the Clean Air Act, establishing the E.P.A. and beginning a nationwide drive to make environmentalism a moral goal, along with a sudden, almost instantaneous skyrocketing of federal environmental laws.[32]

The irony of this profusion of legislation is how little such an approach differed from that of Nikita Khrushchev's. As he believed, "Centralization was the best and most efficient system . . . [Everything] had to be worked out at the top and supervised from above."[33]

* * *

Apollo 8's effect on American society had other short-term consequences as well. "It certainly would not be a very inviting place to live and work," Borman had said while in orbit around the moon. For most of the twentieth century, people had dreamed not only of exploring the moon, but of bringing humanity to it to live. Now the first humans to see the moon up close found it too inhospitable for their taste.

Less than four months before the Apollo 8 mission, the National Academy of Sciences urged NASA to eliminate almost all manned exploration and replace it with unmanned missions. "The ability to carry out scientific observations at a distance is developing so rapidly that I don't see any unique role for man in planetary exploration," noted Gordon MacDonald, chairman of the academy panel that issued the recommendation.[34] While few paid much attention to this recommendation before Apollo 8, afterward the calls to adopt it were many and insistent. For the first time since the nation was founded, respected and powerful voices were saying that sending explorers to open up vast new territories, to take daring and courageous chances for the sake of human advancement, was not in the interests of the United States.

And people listened. It was as if this nation of pioneers had become terrified by what had been shown during those televised broadcasts from the moon, and its citizens no longer wished to accept the challenge of bringing life to a barren world like the moon. Interest in space exploration waned and the space program wound down. When Jim Lovell flew on Apollo 13 sixteen months later, no television network was much interested in covering the mission, until something went wrong. By the late 1970's, the United States essentially had no operating space program, flying no manned missions from 1975 through 1981. In fact, in 1979 NASA launched only three satellites, two small short-term atmospheric research probes and one astronomical X-ray telescope.

Even today, our plans for the human exploration of space are entirely limited to earth orbit. The idea of sending humans to another planet seems hard to fathom. After taking that brave leap to another world thirty years ago, we have become strangely fearful, and are once again hugging the coasts of earth, unwilling to brave the "oceans" that surround us in order to visit the planetary "islands" that exist nearby.

* * *

The vision of the earth given to us by Apollo 8, "a grand oasis in the big vastness of space," did more than merely energize the environmental movement: it encouraged the concept of a single human culture. Before Apollo 8, the earth had always been seen as that far horizon line, a flattened curve beyond which lay alien cultures and foreign lands. Each culture competed, sometimes peaceably, sometimes not, to exert its influence on human history.

After Apollo 8, however, the human vision of our mother planet changed. The image of a "bright, blue marble" in the starkness of the void was far more compelling than anyone had dreamed. The day after the astronauts read from Genesis, the *New York Times* printed a short commentary by poet Archibald MacLeish, in which he gave his interpretation of the Apollo mission.

> To see the earth as it truly is, small and blue and beautiful in that eternal silence where it floats, is to see ourselves as riders on the earth together, brothers on that bright loveliness in the eternal cold — brothers who know now they are truly brothers.[35]

Now the earth was no longer seen as land over which nations could claim control. While borders might exist in the differences and diversity of human culture, the planet itself was one.

Frank Borman's own words over the next few months illustrate this change. A man who had spent his life defending the American idea of freedom could not help but espouse the idea of world peace and cooperation in his

public tour following the mission. "I look forward to the creation of an
international outpost on the moon," he told an audience in Paris.[36] In
Germany he said that space exploration "will teach us that we are first and
foremost not Germans or Russians or Americans but earthmen."[37]

Borman said this with the same cheerful goodwill that had led him
to speak words in lunar orbit to include as many people as possible.
Furthermore, he spoke in the conviction that somehow the Soviet
dictatorship had to be defeated, and perhaps by offering a carrot instead of
a stick he might help in that defeat.

Nonetheless, his words contributed to the idea that there was no
qualitative difference between the free capitalist American vision and the
state-run, communist vision, that the defense of individual freedom was
less important than world peace and world cooperation. In the ensuing
years this idea has grown so strong that it has almost become impermissible
to be proud of our traditions, our unique freedoms and successes.[38] Instead,
cultural pressure insists that we promote the idea that borders are
unimportant and all souls and cultures must be merged into a single global
village, as seen from Apollo 8.[39]

And yet, does this "one world" vision accurately portray our earthbound existence? Or is human experience far more complex?

Less than two months after landing in the Pacific, Frank Borman and
his family were touring Europe as representatives of the U.S. Government.
They visited London, Paris, Belgium, the Netherlands, Spain, and Portugal.
In Rome, Pope Paul VI sat and spoke with them for over an hour.

In West Berlin, Borman found a changed city from his last visit in
1949. The ruins were gone, the poverty and starvation forgotten. Instead he
saw a vibrant city filled with skyscrapers and wealth.

And yet, the Cold War still dominated the view out his hotel window.
Despite the vision of single, unified earth that Apollo 8 portrayed, here was
illustrated a contrast that was as stark and plain as night and day.

One night the Borman family went for dinner in their hotel's
penthouse restaurant. From their table they had a clear view over the city. To
the west was a bright and glittering jewel, the lights of West Berlin twinkling
gaily. To the east was darkness and stillness, the drab streets of East Berlin

cloaked in gloom. And at the dividing line was a barbed wire and concrete wall, lit by guard towers and searchlights.

Despite its appearance as a single globe from the distance of the moon, up close that blue-and-white earth still held upon it some brutal differences.

## CHAPTER FOURTEEN

# THIS
# IS NOW

## FAMILY

ON JANUARY 30, 1998 Jim and Marilyn Lovell joined their son Jay, now forty-two, for the groundbreaking of a new restaurant in the northern Chicago suburb of Lake Forest. Marilyn found the site, Jim financed the construction, and Jay would be the chef.

While Jim Lovell had always known what he wanted to do with his life, Jay Lovell had spent many years searching for something that would infuse his life with a similar joy. After graduating from the Houston Academy of Art he became an illustrator, working first at NASA, and then for the *Houston Chronicle*. Then he tried running his own graphic design company, but found the work unfulfilling, both financially and creatively.

He had always loved cooking, and one day decided he'd like to try doing this for a living. He applied and was accepted to the Culinary School at Kendall College in Evanston, Illinois. To his delight and pleasure he quickly found himself ranked first in his class. Soon he was an executive chef, winning awards at several different high-class restaurants in the Chicago area.

Now he and his father were business partners. Their restaurant, "Lovell's Lake Forest Inn" opened in the fall of 1998, celebrating the thirtieth anniversary of the flight of Apollo 8. In it the Lovell family showcases Jim Lovell's space career and Jay Lovell's cooking talents. "This restaurant will be open for as long as I am alive," says Jay.

For Jim Lovell, the flight of Apollo 8 made remarkably little difference to his life. As soon as the parades and parties ended, Jim returned to the program. *Why stop now, after getting within seventy miles of the moon's surface?* he thought. He became a backup to Apollo 11, which put him in the ideal position to get the assignment as commander for Apollo 13, scheduled to land on the moon sometime in the next two years. He would then become the fifth man to walk on the surface of another world.

Unfortunately, halfway to the moon Apollo 13's oxygen tank exploded. To get the men home alive and unharmed demanded tremendous patience and technical skill, both by the astronauts themselves and their colleagues at NASA.

This meant, however, that Jim Lovell would never walk on the moon. In the 1970's the American space program was dying, and Jim could see that it might be years, if ever, that he would fly into space again. He decided it was time to leave NASA and try his hand in private enterprise. He went looking for something new to do with his life.

For four years he was president of a tugboat business in the Houston area. Then he became president of a telecommunications firm, selling business phone systems in the Southwest. Later, he was executive vice president for Cintol Corporation, which had purchased his Houston company. He and Marilyn moved back to the Midwest where they both had grown up, and built a home in the suburbs north of Chicago. As the years passed, the Lovells faded into the pleasant but obscure life of middle America. Once in a while a reporter would call to ask them about Jim's Apollo missions, but in general it seemed the country had lost interest in space. Jim and Marilyn watched their children grow up and go out on their own, becoming ordinary Americans in the never-ending, always changing American landscape.

Finally, Jim decided to sit down with writer Jeffrey Kluger and write *Lost Moon: The Perilous Voyage of Apollo 13*. Even if he couldn't fly again,

Credit: Lovell

The Lovell family, 1997. Jim and Marilyn Lovell are in the back row,
on the left. Barbara Lovell stands in front of Marilyn, with Jay
Lovell to the right in the center of the picture. Susan Lovell
is in front of Jay wearing vest, and Jeffrey Lovell is directly
behind Susan in the black shirt.

perhaps in telling a new generation the thrilling story of his last flight to the
moon he could generate some new interest in space exploration.

The book became a movie, *Apollo 13,* and eventually sold over a
million copies. Though Jim Lovell would certainly not take full credit for the
recent renewal in American space exploration, he surely has the right to claim
some of the accolades.

\* \* \*

For Lovell, the flight of Apollo 8 did little to change his religious beliefs. "Going to the moon to me was not a religious event," he explains. To him, if you believed in God you could find Him anywhere, either on earth or in a tiny capsule floating in the darkness of space. Two hundred forty thousand miles was "just a drop in the bucket" for an entity that had created the entire universe.

To Bill Anders, however, his journey to the moon radically changed his outlook on life and religion. Ironically, while the world was celebrating the astronauts' affirmation of spirituality, the flight had undercut Anders' own faith in Catholicism. The vast emptiness of space made the Catholic rituals he had obeyed faithfully since childhood seem insufficient to him. "We're like ants on a log," he explained. "How could any earth-centered religious ritual know what God's truth is?"

His doubt made it impossible for him to perform these rituals with his same past sincerity. Consequently, he simply stopped doing them. He ceased attending church.

This change caused some family problems, though fewer than might have been expected. Valerie herself had had increasing doubts about their religion, even before the flight. She disagreed with the Catholic Church's opposition to contraception. Even more disturbing to her was her knowledge of women who had been able to shop around and find a Catholic priest who would sanction their use of birth control pills. Such arrangements made Valerie doubt the Church's sincerity.

Yet she also wondered how she as a parent could raise her children without some greater guiding principles for explaining right and wrong. "Without religion you lose the security of a church helping you raise a child. You're left on your own, trying to explain a lot of difficult ideas without support."

Valerie worried that Bill's abandonment of religion might alter the sanctity of their marriage. They had been married in accordance with the Church's rules, "until death do us part" as it were. Would those rules still hold if they no longer believed in Catholicism?

They talked about it at length. Finally they decided that they could still lead an upright, virtuous life, even if they did so without the Catholic Church. Bill would honor their marriage, regardless, Valerie knew. As

The Anders family 1995. Left to right, Glen Anders is second on the left, standing with Valerie to his right holding grandchild. To her right wearing white hat is Greg Anders. Diana Anders wears ten gallon hat in center. Bill Anders, in white hat and vest stands with Alan Anders to his left behind him and Gayle Anders to his right wearing white hat. Eric Anders is on far right.

astronaut Walt Cunningham had written, Anders belonged (as did Frank Borman) to the "end of the spectrum [of] guys so straight that you didn't know whether to admire them or have them stuffed and shipped to the Smithsonian Institution as the last of a vanishing species."[1]

When their last child, Diana, was born in 1972, they raised her outside any organized religion, dealing with the problems this caused as they came up. "Diana would wonder why her friends were going to church, and we weren't," Valerie explained. She told her daughter that just as Valerie's parents had allowed her to choose her faith, the Anderses were allowing Diana to do the same.

While Bill Anders very much wanted to walk on the moon like Jim Lovell, the likelihood of his doing so was slimmer. After Apollo 8 he was assigned to the position of command module pilot, working backup for Apollo 11 and prime crew on Apollo 13. This assignment meant, however, that the best he could hope for was to watch the others go down and land while he remained in lunar orbit. It didn't seem worth the risk to go back to the moon, merely to go around it again.

Furthermore, President Nixon had offered him a position in government as executive secretary of the National Aeronautics and Space Council. There he would have the ability to directly influence future space policy. Anders took it, and before too long was immersed in the political world of Washington. In 1973 he was appointed a member of the Atomic Energy Commission, and shortly thereafter became the first chairman of the Nuclear Regulatory Commission. Later he served as ambassador to Norway.

Then he entered the private sector, working for a variety of companies in the military and airplane construction industry. Eventually he moved up the corporate ladder to become C.E.O. of General Dynamics, trimming that company's overhead and selling off many of its non-profitable divisions.

Today Bill Anders is retired, and spends his free time flying airplanes for fun. And if asked, he would go to the moon in a heartbeat. Indeed, he has applied to NASA to follow John Glenn back into orbit.

* * *

Frank Borman, however, has absolutely no interest in going back in space. After the flight of Apollo 8 he was quite content to do work that was less dangerous and put less stress on his wife.

For a while he flirted with the idea of joining the political world, going on a trip to the Soviet Union in conjunction with President Nixon's policy of detente. Then he decided, like the other two astronauts, that the challenges of the corporate world were more interesting. He joined Eastern Airlines' management team, and eventually took over as its C.E.O., trying to shepherd that company through the de-regulation of the airline industry.

It was the only challenge in his life Frank Borman failed to meet. Eastern's union wages were too high. For four years he managed to get the unions to go along with reduced wages and profit-sharing, while simultaneously straightening out the company's finances. When deregulation hit the industry in 1978, however, one union backed out of the deal, causing the

Credit: Borman

Susan and Frank Borman, about to go flying together, 1997.

whole profit-sharing arrangement to collapse. When that same union next made demands that exceeded the ability of the airline to pay, Borman was faced with either acquiescing to the demands or going bankrupt during a strike. He chose to avoid a strike, hoping he could get the union to reconsider.

The union refused, and in fact increased its demands. Eventually, Eastern was sold, and Borman was forced to step down. On the day he lost the company, this tough no-nonsense test pilot came home and cried in his wife's arms.

But a worse upheaval had earlier struck the Borman family. By the early 1970's both Fred and Ed were attending West Point, and Susan Borman found herself for the first time alone at home without a family to raise. The doubts and fears that she had experienced during Apollo 8 had continued to haunt her, and she increasingly resented her earlier willingness to bury herself entirely in the life of her husband. Even though he no longer had a life-threatening occupation, he was still obsessed with it, and put in endless hours at Eastern. With her two children now grown, Susan felt trapped by Frank Borman's life.

Frank Borman with grandchildren, 1983.

She began to drink. She eventually had a nervous breakdown. For a time Susan hated her husband, her life, and her existence. Similarly, Frank felt a terrible remorse and shame for always putting Susan second and his career first. After a period of therapy and intense soul-searching, however, they began rebuilding their lives together.

Today, they live in Las Cruces, New Mexico, where Frank owns and operates a major car dealership as well as a number of other businesses. No longer does he do work that could get him killed. No longer does he work so hard that he ignores the needs of his wife.

Susan, meanwhile, has learned to live her own life, without abandoning her deep abiding love for Frank. And she has become significantly more religious.

Even now, thirty years after Frank's return from the moon, she still cannot believe that the mission succeeded. To her, the odds had been too great. Divine intervention must have played its part. As she told me, "I honestly to this day believe, with every cell in my body, that [Apollo 8] was a miracle."

*  *  *

In the end, the thought that defines these three families is that of a promise kept and an oath fulfilled. These six men and women accepted their oath of marriage as a bond to be honored, making a lie of the modern myth that enduring marriages are impossible. These were partnerships for life, and each partner had a task to make the marriage work. In the end, everything they did, they did for each other.

## EARTH

Fred Gregory held on for dear life. Once again his body was shaking like crazy, but unlike his helicopter experience in Vietnam, this time he knew what was happening. Below him roared a spaceship weighing four-and-a-half million pounds, its engines firing almost six-and-a-half million pounds of thrust. In mere minutes he was traveling more than a dozen times faster than he had ever flown in a jet.

Fred Gregory, a man who had had absolutely no interest in space exploration when Borman, Lovell, and Anders had gone to the moon, was now on his way into earth orbit. The date was April 29, 1985, and Fred was pilot of the space shuttle *Challenger*.

Much had happened since he had flown that helicopter in the jungles of Vietnam. By 1973 the United States's armies had withdrawn, after more than 50,000 American and many more Vietnamese casualties. By 1975, North Vietnam had overrun the South and won the war.

Gregory finished his tour of duty in 1967 and came home looking for new and challenging aviation work. He went to the library to research what jobs were available to skilled military pilots, and learned that there were military schools where he could train to become a test pilot, flying experimental jets in the most dangerous circumstances. He applied, was accepted, and completed the test pilot course at the Naval Test Pilot School at the Patuxent River Naval Air Station.

By 1970 he was an engineering test pilot at Wright-Patterson Air Force Base, flying both helicopters and jets in the craziest of situations. For example, Fred cheerfully volunteered to pilot a jet into thunderstorms so that meteorologists could study their violent and deadly wind and electric patterns. He did this more than fifty times, letting the winds buffet and shake his airplane so hard that many times it was almost torn apart.

When NASA announced in 1976 that it was looking for pilots for its new space shuttle, Fred was intrigued by the idea of flying this new kind of radical "airplane." Unlike the space capsules of the 1960's, this looked like something he could pilot.

Almost a decade after Apollo 8 he still knew little and cared less about the space program. In fact, though his wife Barbara insists that they watched the reading of Genesis in his parents' home on Christmas Eve, 1968, Gregory today doesn't remember this moment at all.

His interest in the shuttle program and the exploration of space finally came alive one day when he was home watching television. Nichelle Nichols, who had played Lt. Uhura on the science fiction television show *Star Trek*, was spokeswoman on a NASA public service announcement calling for new astronaut applicants. She looked into the camera and said, "We want *you!*"

Fred looked back at her in shock. He felt almost as if she were talking directly to him. He sent in his application, and by 1978 he was in the program.

Now it was 1985, and Fred Gregory was flying his first mission in space. Once in orbit, the crew settled down to routine business. This mission was the second Spacelab mission to fly, and in the cargo hold a crew of five scientists conducted experiments on materials processing, atmospheric physics, astronomy, fluid behavior and life sciences.

As pilot, however, Gregory had little to do with the experiments. Instead, he and his commander, Robert Overmyer, alternated twelve hour shifts, sitting alone in the cockpit, monitoring the shuttle's performance as it orbited the earth time after time.

For seven days Fred circled the earth. Though he didn't know it, his experience was remarkably similar to that of Borman, Lovell and Anders. At one point the parallels were uncanny. Though the mission itself went smoothly, the life sciences experiments involved two monkeys and two dozen white rats. Unfortunately, the cages were not well sealed, and before too long floating animal feces and dried food particles drifted throughout the shuttle. Like the Apollo 8 astronauts decades before, Fred spent a good portion of his free time trying to scoop floating feces out of the air.

In other, more significant ways, Gregory's experience was very different from that of the astronauts on Apollo 8. The shuttle was large, more closely resembling the spaceships in the movie *2001: A Space Odyssey* than the Apollo 8 capsule. The shuttle had three separate decks, and could accomodate eight astronauts comfortably, and ten in an emergency. Its atmosphere was a mixture of twenty percent oxygen and eighty percent nitrogen, rather than pure oxygen. The kitchen area on the middeck included an oven for reheating food packages, color-coded utensils and food packs for each astronaut, and condiments such as salt, pepper, taco sauce, hot pepper sauce, catsup, mayonnaise and mustard. Its cargo bay was so large that four Apollo 8 capsules could fit within it. Rather than splashing down in the ocean, the shuttle landed on a runway like an airplane. And above all, it was *reusable.*

These were the engineering differences. Other disparities were far more profound. When Gregory stared down at the blue, white, and brown planet below him, he did not see a fragile, delicate Christmas ornament as did Borman, Lovell, and Anders. Instead, he saw something incredibly robust and sturdy, a grand and tough planet that actually seemed capable of healing itself repeatedly despite any and all forms of injury.

This impression was further strengthened on Gregory's final shuttle flight before he retired from active duty. On November 25, 1991 he took off as commander of the shuttle *Atlantis* on a seven-day mission. Besides putting a military surveillance satellite into orbit, the crew spent most of its time conducting earth observation experiments.

In 1991, the earth's atmosphere had undergone a number of violent upheavals. In January the Iraqi military had fled Kuwait, igniting 732 oil wells as they left. For months these wells burned, sending tons of smoke into the air. Some experts worried that the black clouds could affect the earth's climate. When *Atlantis* reached orbit in late November the last well fire had been quenched only two weeks earlier.[2] Then, in June, the Mount Pinatubo volcano exploded, sending columns of ash and smoke twenty-five miles into the air, forcing the evacuation of hundreds of thousands of people, destroying whole cities, and dumping over two cubic miles of volcanic ash across a million-and-a-half square mile area from Indonesia to Vietnam.[3] Within three weeks Pinatubo's plume had completely encircled the globe, spanning fifty degrees of latitude worldwide.[4]

What amazed Gregory was how little visible evidence remained from earth orbit of these ferocious events. Only five months after Pinatubo's eruption he could see little atmospheric evidence of that gigantic and furious explosion. Though he knew intellectually that the eruption's giant aerosol cloud still permeated the atmosphere worldwide, he could not see it.

Even more surprising was what he saw of the Kuwaiti oil fires. Though Gregory could still see some evidence of the fires' smoke in the atmosphere above the Middle East, he was amazed to see it diminish each day. By the time he returned to earth the air actually seemed clearer. "I equated [the earth] to a cat cleaning itself," Fred told me. "When it became dirty it just licked itself clean."

The Apollo astronauts had looked at the earth and had seen a tiny, frail object in the blackness of space. The Apollo astronauts, however, were flying in a tiny, very frail capsule, voyaging far beyond the utmost limits of human capability. Unconsciously, they were projecting their own fragility back on to a distance earth.

Fred Gregory, however, floated in his shirtsleeves in a true spaceship, as hardy and as safe as any combat jet. He did not feel himself constantly vulnerable to certain death, mere inches away. Nor did his experience push the envelope of human experience farther than was prudent or safe. Instead, he was doing what humans now considered risky but ordinary work, though

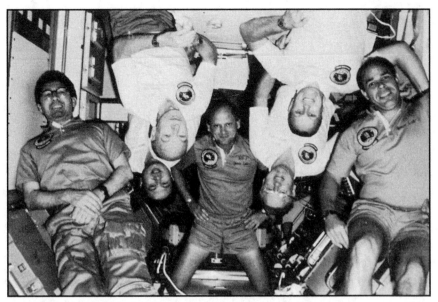

The crew of shuttle mission 51-B floats in orbit, May, 1985.
Fred Gregory is on the far right.

that work did extend the human experience outward into regions that were little explored and newly won.

Hence, Fred could look down at the earth and see another vision, that of a large *planet* with a complex ecology, able to balance and maintain itself, despite terrible afflictions. Long after every human had died, Gregory knew that that planet would go on renewing itself.

That Gregory's perspective was different from the astronauts of thirty years ago is hardly surprising. While his early lack of interest in space exploration might have reflected the lean years of the 1970's — when few in America were interested in exploration — his desire to go into space in the 1980's was a clear precursor of the space boom happening this very moment. Not only has NASA sent probes to Mars, the moon, and Saturn for the first time in decades, the first components of the American-led international space

The clear blue Pacific Ocean, west of the Philippines, as seen by Fred Gregory, 1991.
The light haze in the center of the picture is the sun's reflection.

station have been completed, and should Russian funding problems finally be solved, it will reach orbit sometime within the next year.

More significant is the boom in private space development. Revenues from commercial space launches in 1997 totaled $85 billion, and are expected to increase to $121 billion by the year 2000.[5] One company has launched a constellation of seventy-two satellites in its plan to provide cellular telephone service anywhere in the world.[6] Another has launched eight out of a planned array of forty satellites, and a third has launched twelve of thirty-six. And in 1997, commercial launches exceeded government launches for the first time in history.

Today, private enterprise dominates the space industry. Several industry studies predict that in the next decade between 1,700 to 2,000 new satellites will be launched, with seventy percent of these commercially financed.[7] With this many satellites planned, a gigantic need has developed for new and cheaper rockets. Almost a dozen private firms are developing reusable spaceships.[8] One man, Andrew Beal, has invested a quarter of a billion dollars of his own money to develop a new expendable rocket.[9] Another businessman, Jim Benson, has formed a corporation to send an unmanned mission to a nearby asteroid, with launch expected sometime prior to February 2001. Though some might think this plan farfetched, Benson's company, SpaceDev, estimates sales in its aerospace and engineering divisions will exceed ten million dollars in 1998. According to the company's literature, "SpaceDev believes there is pent-up demand for economical space exploration."

The new century will see a renaissance of space exploration as exciting and as challenging as the space race of the 1960's. And this rebirth will happen under the banner of freedom and private property, the very principles for which the United States fought the Cold War.


## FREEDOM


And in Berlin, the wall is no more. Where that death strip of barbed wire and guard towers once stood are gleaming office buildings and shopping malls. No longer do foreign troops patrol the city. The only indication that remains of the forty year head-to-head conflict between capitalism and communism is a small museum one block from Checkpoint Charlie (where a giant office complex now stands). In this museum are many of the tools and equipment used by the thousands who succeeded in vaulting the Berlin Wall to freedom. Also there are testimonies to the over eight hundred people thought to have died trying to escape East Germany.[10]

The Cold War is over, and the Soviet Union is a memory. As Ronald Reagan correctly predicted in 1982: "The march of freedom and democracy . . . will leave Marxism-Leninism on the ash heap of history."[11]

Despite Khrushchev's claim that Nixon's grandchildren would live as communists, it is Nikita's descendants who today live as capitalists, striving hard to catch up with America after more than seventy years of communist rule.

Perhaps Frank Borman expressed it best. In summing up the fall of communism he said very simply, "In the final analysis, everybody wants to be *free*."

Ironically, possibly the only good legacy left by the communist dictatorship was its space program. For more than twelve years, space station *Mir* has dominated space exploration, providing a platform for science and international prestige. Even as the Soviet Union collapsed and disappeared from the earth, an abject demonstration of the failure of a centralized state-run society, its space station has lived on.

Today, the Russian government owns *Mir*, and clearly recognizes its value. In fact, the Russians have taken a decidedly capitalist approach to maintaining their entire space program. Forced to raise cash, the Russians have eagerly sold as advertisement space the walls of their mission control, much like a sports stadium. They have commercialized their launch services, offering them to private communication satellite companies.

And they have rented their space station to such Western countries as England, France, Germany, and even the United States. In exchange for substantial cash payments in the hundreds of millions of dollars, foreign astronauts have visited *Mir* and used it for training and scientific research.

In America, meanwhile, ordinary Americans still do what they have always done, despite the cultural pressure to deny the existence of a distinct American way of life. When astronaut Michael Foale returned to earth after spending four months on *Mir*, he noted that "I have, of course, thought a lot about my family . . . And my priority now is to spend more time with my young children over the next year or two . . . I'm looking forward to the adventure of learning how to walk again and live in my house with my wife and my children, get to know my wife again, date her again, maybe marry her again."[12]

And Foale's replacement, David Wolf, told reporters prior to launch that while in space he would observe the Jewish Day of Atonement, Yom Kippur, including fasting and prayer. Unable to attend synagogue services, Wolf pointed out that he could "still do it in mind."[13] Furthermore, during his four month stay, Wolf became the first human to vote from space, casting his ballot for local elections in Houston. A new law in Texas had made it possible for an astronaut in space to do this electronically.

Jim Benson, meanwhile, in his commercial plans to explore an asteroid, intends "to make an ownership claim of the asteroid, [setting] a precedent for private property rights in space."[14] As Benson stated recently, "Any attempt by world legal bodies to limit such property rights in space will . . . be viewed not only as a 'taking' but as a threat to anyone who has any interest in going to space for work or play, or who might have a job on earth directly or indirectly related to space commercialization."[15]

As always, family, freedom, and moral commitment remain deeply engrained in the American mind. These ideas permeate everything we say and do, just as they permeated everything the Apollo 8 astronauts had said and done. And such ideas still strengthen the ability of individual Americans to fulfill their dreams, whatever those dreams might be.

As we enter the third millennium, the human race will at last embark on the permanent exploration and settlement of space. No longer will we hug the coast, fearful of the vast black ocean between the planets. No longer will we see the earth as our only safe haven in a dangerous universe. Instead, we will see it as Fred Gregory does, a beautiful blueprint for the noble task of bringing the earth's vibrancy to other worlds, to make the moon blossom like a garden, to bring Mars back to life, to end the sulphuric storms of Venus and allow children to play on its windswept volcanic shores.

When we go, we should also bring with us a good blueprint for human society. Like Jim Benson, we should, as free men and women, bring with us the laws of the United States and the capitalistic and democratic principles of our country. And like Mike Foale and David Wolf, we should also infuse the future generations of space settlers with principles of family, freedom, and moral commitment.

Kennedy had said we must. And we should, not for nationalistic reasons, but because we as a nation and culture have stumbled upon a good formula for human society. We aren't better than anyone else, and surely have many faults and weaknesses. And though we have made our share of evil decisions in our history, far more often we have done right for ourselves and for others.

As surely as the sun has risen this morning, and as surely as it will set this evening, the human race is going to the stars. Though we might do it for the adventure and profit and the search for knowledge, we will eventually settle down on those barren hills that Frank Borman called "not a very inviting place to live and work." And when we do, we had better consider the social order that we establish. The Cold War might be over, but it was only a small episode in the never-ending struggle between freedom and tyranny, a battle that began when the first bully found he could use a club to make others do what he wanted.

The Pilgrims came to the New World to escape just such a bully. And though they made their own transgressions in time, they managed to consciously and carefully establish a good social order, leading to some spectacular and glorious results.

We should do no less when we reach for our own new worlds, out there amid the stars.

# NOTES

## INTRODUCTION

1. Hoyle, pp. 9-10.
2. *National Geographic,* 2/98, 44
3. Gagarin, 193
4. Julian Scheer interview, 10/23/97

## CHAPTER ONE

1. Clarke, 161
2. *Orlando Evening Star,* 12/21/68
3. Chaikin, 116
4. *La Mesa Scout,* 1/2/68, 1
5. Borman, 197-198
6. Borman, 297; Aldrin, 123
7. Both Andrew Chaikin's *A Man on the Moon* and the hardback edition of Lovell's *Lost Moon* describe how Susan expressed anger at Kraft because of NASA's decision to send Apollo 8 to the moon. According to Susan Borman, this is incorrect. "I was never angry," she explained to me. She told Kraft how she was sure the men would die in lunar orbit, and he tried (and failed) to reassure her.
8. Borman, 66
9. Borman, 303-304
10. *Milwaukee Journal,* 2/21/69
11. Borman, 196; Chaikin, 79; *Life* 2/69, 19
12. Borman, 196-197; *Life* 2/69, 19
13. *Life,* 2/28/69, 19
14. *Washington Post,* 12/22/68, 1; *Houston Chronicle* 12/21/68, 1
15. Baker, 292-294; Lay, fronts piece; Borman, 133
16. Baker, 294
17. Murray and Cox, 242-250; Chaikin, 86
18. Borman, 201
19. Chaikin, 87

20. Collins (1989), 288-300
21. Borman, 203
22. Borman, 204

## CHAPTER TWO

1. Gun, 268-270, 299-303
2. *New York Times,* 1/20/46, 25; Gun, 268-270, 299-303
3. *New York Times,* 7/14/47, 7
4. Borman, 35
5. Churchill, 359-362, 373-374, 400-407
6. Davison, 8-10, 13-18, 27-30, 144-149
7. Davison, 93
8. Collier (1978), 59-83
9. Collier (1978), 186
10. O'Ballance, 118
11. O'Ballance, 126-128, 130-135; Kousoulas, 164-165, 177-178
12. O'Ballance, 191-202
13. Lovell (1952), 24
14. Lovell (1994), 66-72
15. *Newsweek,* 2/3/69, 11; *Washington Post,* 8/31/69, 2
16. *New York Times,* 7/25/59, 3
17. *New York Times,* 1/25/59, 1; Nixon (1978), 206
18. Khrushchev (1970), 13-35; Khrushchev (1990), 1-13; Medvedev, 3-24; Pistrak, 3-14
19. Khrushchev (1970), 21-22; Medvedev, 13, 18-19; Pistrak, 15-18, 30-32
20. Khrushchev (1970), 36-119; Medvedev, 25-39; Pistrak, 27-85
21. Whitney, 1-7
22. *New York Times,* 11/18-20/56, 11/20/56, p15
23. Burlatsky, 76-78, 132-136
24. Khrushchev (1970), 385
25. Khrushchev (1974), 79
26. Shlapentokh, 105-109; 149-151; Gunther, 55-57, 284-288; Brumberg, 360-390; Johnson, 1-89
27. Anderson, 19
28. Anderson, 19
29. Anderson, 1-67; Bourdeaux (1968), 12-19; Bourdeaux (1970), 38-59, 65-84, 97-155, 304-329; Fletcher, 230-272; Corley, 184-243; Struve, 267-335
30. Cate, 27; Gelb, 39-44; Dulles, 41; *Survey* v44-45, 10/62, 59-65
31. Khrushchev (1990), 167
32. Khrushchev (1990), 120-127; Burlatsky, 88-92
33. Whitney, 5
34. *New York Times,* 6/30/59, 1, 17; 7/2/59, 1; Barghoorn, 12-13, 92-94, 134
35. Shepard & Slayton, 42
36. Breuer, 150
37. *New York Times,* 6/30/59, 16

38. Chaikin, 39
39. Beschloss, 701-706


## CHAPTER THREE

1. Baker, 250-255, plate opposite 273
2. Borman, 203
3. Collins (1989), 60; Cunningham 204
4. Borman, 194
5. Hall, 122
6. *Houston Post,* 12/22/68, 1
7. Murray, 326; Phillips, 604-607
8. Phillips, 642-643
9. *Houston Post,* 12/22/68, 1, 25
10. *Milwaukee Journal,* 12/23/68, 1:6
11. *Milwaukee Journal,* 12/23/68, 1:6
12. Armbrister, 33-78, 224-244; Murphy, 120-203
13. *New York Times,* 12/23/68, 1


## CHAPTER FOUR

1. Cate, 250
2. Schick, 162-163
3. Gelb, 63
4. Khrushchev, (1990), 169
5. Bourdeaux (1970). 65-84.
6. Khrushchev (1974), 506
7. Khrushchev (1974), 505
8. *New York Times,* 8/25/61, 5
9. Cate, 304
10. Beschloss, 114-134, 143-147; *New York Times,* 4/16/61, 1; 4/18/61, 1; 4/20/61, 1
11. Gelb, 74
12. *New York Times* 2/22/61, 1; 3/19/61, 42; Baker, 63-65
13. *New York Times* 4/13/61, 34
14. *New York Times* 4/13/61 14
15. *New York Times,* 5/26/61, 12
16. *New York Times,* 5/26/61, 1, 12-13; Logsdon, 64-130; Lambright, 82-101; Beschloss, 165-167
17. Lovell (1994), 180-181
18. *New York Times,* 8/16/62, 10
19. *New York Times,* 8/16/62, 26
20. Galante, 170
21. Galante, 178-186; Ausland, 59-63
22. *New York Times,* 8/18/62, 1

23. *New York Times*, 8/19/62, 7
24. *New York Times*, 8/17/62, 4


## CHAPTER FIVE


1. *Life*, 1/10/68, 80-81


## CHAPTER SIX


1. Shepard and Slayton, 172.
2. Orberg, 80-82; Aldrin, 122-123.
3. Shepard and Slayton, 174.
4. Orberg, 75-79; Baker, 188; Aldrin, 110.
5. http://solar.rtd.utk.edu/mwade/flights/voskhod1.htm
6. *New York Times*, 10/13/64, 18.
7. *New York Times*, 10/17/64 12.
8. Aldrin, 120; Baker, 188-190.
9. Borman, 133.
10. *New York Times*, 12/7/65, 21.
11. Borman, 137.
12. Borman, 137.
13. *New York Times*, 12/10/65, 22.
14. Borman, 139.
15. *New York Times*, 12/13/65, 1, 46; Baker, 219-225.
16. Borman, 141.
17. Borman, 129.
18. *New York Times*, 12/14/65, 24; Baker, 224.
19. *New York Times*, 12/15/65, 1, 28.
20. *New York Times* 12/17/65, 28; Borman, 144.
21. *New York Times*, 12/19/65, 68.
22. *New York Times*, 12/19/65 69.
23. *New York Times*, 12/19/65, 69.
24. *New York Times*, 12/19/65, 68.
25. *New York Times*, 8/5/64, 1.
26. *New York Times*, 8/6/64, 8.
27. *New York Times*, 12/20/65, 6.
28. Shadrake, 67-71, 77-79; Galante, 150-152, 167-170.
29. Shadrake, 112-123.
30. *New York Times*, 1/12/66, 12.
31. *New York Times*, 2/3/66, 13; 2/4/66, 8.
32. *New York Times*, 2/2/66, 3; Shadrake, 102-105.
33. *New York Times*, 12/22/65, 7; 12/23/65, 15; 12/26/65, 72; 12/27/65, 3.
34. *New York Times*, 12/5/65, 1.
35. *New York Times*, 11/21/65, 60; 11/28/65, 1; 11/29/65, 3.

36. *New York Times,* 8/12/65, 15; 8/13/65, 1; 8/14/65, 1, 8; 8//16/65, 18; 8/17/65, 1.


## CHAPTER SEVEN

1. Guiley, 14, 31, 47-48, 50-54
2. Verne, 634
3. *New York Times,* 12/24/68, 6
4. *New York Times,* 12/25/68, 38
5. Both the *New York Times* (12/25/68, 38) and the transcripts (02 21 51 16) say that Lovell said "deep sand," but he, Bill Anders, and Thomas Paine all remember the phrase as "beach sand."


## CHAPTER EIGHT

1. Baker, 276-279; Aldrin, 161-165.
2. Lovell (1994), 189-192.
3. Borman, 170.
4. Borman, 173.
5. *New York Times,* 4/23/67, 56.
6. http://solar.rtd.utk.edu/mwade/flights/soyuz1.htm
7. Lebedev, 161-168.
8. *New York Times,* 4/7/68, 1, 62, 63.
9. Cox, 73-74; Kahn, 108.
10. *New York Times,* 4/12/68, 1.
11. *New York Times,* 1/2/68, 4; 2/7/68, 15.
12. *New York Times,* 2/4/68, 11; 2/4/68, 15; 2/9/68, 1, 12.
13. *New York Times,* 2/6/68, 16; 3/6/68, 1.
14. *New York Times,* 4/26/68, 3.
15. Adelson, 124-130, 142-145, 206-208; Kelman, 107-161; Heath, 56-71, 118; Unger, 117-148; See also *New Left Notes,* 1966-1971.
16. Kahn, 76.
17. Kahn, 182.
18. Cox, 206.
19. Kahn, 98.
20. *New York Times,* 5/13/68, 46; Cox, 59, 205-207.
21. Kahn, 127; Unger, 107; Adelson, 223.
22. Adelson, 223-224
23. Kahn, 150
24. Kahn, 151
25. Kahn, 183
26. Hayden (1988), 272-282
27. Cox, 156-168; Kahn, 188-213; Baker, Brewer, DeBuse, Hillsman, Milner, and Soeiro, 29-76
28. *New York Times,* 5/18/68, 1; 5/19/68, 1

29. *New York Times,* 5/22/68, 1; 5/23/68, 1; Baker, Brewer, DeBuse, Hillsman, Milner, Soeiro, 93-95
30. *New York Times,* 5/12/68, 69
31. *New York Times,* 4/27/68, 1
32. *New York Times,* 5/21/68, 1, 51
33. Murray and Cox, 309-312
34. Murray and Cox, 312
35. *New York Times,* 4/8/68, 16
36. Murray and Cox, 315

## CHAPTER NINE

1. Apollo 8 Technical Debriefing, 38-39.
2. The transcripts (03 00 05 40) quote Lovell as saying "I can see the old second bishop right now." This however makes no sense to him. Since Mount Marilyn was "the initial point," the first landmark used by Armstrong and Aldrin as they began their descent to the lunar surface, he believes that this is what he said, and that the transcription is in error.
3. *Parade,* 2/23/69, 20-21.
4. *Discover,* 7/94, 40.
5. Melosh, v, 3-13.
6. Borman, 295.
7. Borman, 207.

## CHAPTER TEN

1. *New York Times,* 2/7/68, 17.
2. *New York Times,* 8/12/68, 1, 15; *Miami Herald,* 12/26/68, 6c.
3. *New York Times,* 8/13/68, 10.
4. Hansen, 245.
5. Aldrin, 191.
6. Murray and Cox, 313.
7. Lay, 140-147.
8. *Daily Mail,* 1/16/69.
9. Chaikin, 66.
10. Lovell (1994), 38.
11. Chaikin, 59.
12. Murray and Cox, 322.
13. Murray and Cox, 322-323; Lambright, 200-205.
14. Pehe, 194-198; Dawisha, 138-141, 148-152; Willaims, 63-111.
15. Williams, 112, 125-143; Dawisha, 319-332; *New York Times,* 8/30/68, 1; 8/31/68, 1-2; 9/14/68, 1.
16. *New York Times,* 5/24/68, 1; 5/25/68, 1; 12/4/68, 1; 12/6/68, 1; 12/7/68, 4.
17. *New York Times,* 6/25/68, 1, 29.

18. *New York Times*, 8/11/68, IV-1.
19. *New York Times*, 11/8/62, 18.
20. *New York Times*, 9/15/68, 1.
21. *New York Times*, 8/30/68, 17.
22. *New York Times*, 8/29/68, 1; 8/30/68, 1; New York Times, 9/1/68, 1.
23. *New York Times*, 11/8/68, 1.
24. *New York Times*, 11/6/68, 3.
25. *New York Times*, 8/30/68, 14; 8/31/68, 11; 12/4/68, 29.
26. *New York Times* 11/8/68, 16.
27. Borman, 183-185.
28. Lovell (1994), 32-33.
29. Baker, 309.
30. Baker, 309.
31. Aldrin, 194.
32. Aldrin, 197.
33. Lambright, 199.
34. Borman, 194.
35. Unfortunately, President Johnson's desire to have a big dinner party with the astronauts on December 9th defeated this plan. Valerie Anders remembered that all she could hear that night at the White House dinner were people coughing. Fortunately no one got sick, though some to this day wonder if Frank Borman's ailment on day one of the mission might have been a twenty-four hour virus.
36. Sherrod interview of Laitin, 3.

## CHAPTER ELEVEN

1. Foerster, 12.
2. Anderson (1993), 145-157.
3. Morgan, 92-93.

## CHAPTER TWELVE

1. *New York Times*, 12/26/68, 1.
2. *Washington Post*, 12/26/68, F1.
3. *New York Times*, 12/26/68, 41.
4. *New York Times*, 12/26/68, 41.
5. Borman, 216.
6. Borman, 216.
7. *Miami Herald*, 12/27/68, 31-A.
8. Borman, 192.
9. Borman, 217.
10. Borman, 218.

## CHAPTER THIRTEEN

1. Borman, 220.
2. *New York Times*, 4/5/69, 19.
3. *Time*, 1/7/80, 3.
4. *Time*, 1/3/69, 17.
5. *Miami Herald*, 12/28/68, 16-A.
6. *Houston Post*, 3/8/69, 14.
7. *New York Times*, 12/8/69.
8. *New York Times*, 12/24/68, 22.
9. Oberg, 111-127.
10. Kamanin, 4-5.
11. *New York Times*, 1/19/69, II-28; 2/2/69, II-31; 2/28/69, 4; 3/16/69, II-38.
12. *New York Times*, 1/10/69, 1.
13. *New York Times*, 12/30/68, 1.
14. Aldrin, 239-240.
15. Chaikin, 204.
16. http://www.hq.nasa.gov/office/pao/History/alsj/frame.html; click on "Post-Landing Activities" and go to 105:26:08.
17. Chaikin, 204-206.
18. Frank Borman notes in his book *Countdown*, pp. 237-238, that Nixon rejected remarks proposed by NASA that would have made it sound as if his "administration was responsible for Apollo 11's success and most of the technological achievements that preceded the flights." Instead he said words that were "appropriate, brief, and [didn't upstage] the Apollo 11 crew." Nonetheless, Nixon did speak, setting a precedent that has since turned almost every space mission into a political photo-op.
19. *New York Times*, 3/18/69, 44.
20. *New York Times*, 5/20/69, 1.
21. *New York Times*, 12/28/68, 47.
22. "People's Science" in *Science for the People*, February, 1971, 3:1, 12, 14-20.
23. Borman, 236.
24. Borman, 234-236.
25. Gitlin, 431.
26. Gitlin, 422.
27. Sinclair, 51-68; Jacobson, 37-42, 205-206.
28. *Sierra*, 9/95, 73.
29. *Whole Earth Catalog*, Menlo Park, California: Nowells Publications, 1971. The caption was a quote from *Energy Flow in Biology* by Harold Morowitz, New York: Academic Press.
30. *New Left Notes*, 2/5/70.
31. *Co-Evolution Quarterly, the Ongoing Whole Earth Catalog*, 6 (1975, Summer), 28.
32. Aronowitz, 163-168. The increase in environmental legislation almost immediately after Apollo 8 can also be seen in the following table, researched by Richard E. Balzhiser. See Balzhiser, 118.

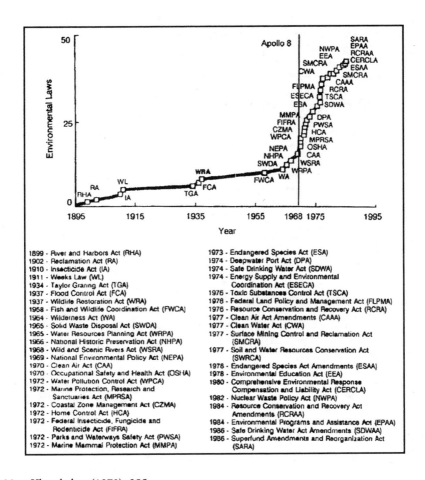

1899 - River and Harbors Act (RHA)
1902 - Reclamation Act (RA)
1910 - Insecticide Act (IA)
1911 - Weeks Law (WL)
1934 - Taylor Grazing Act (TGA)
1937 - Flood Control Act (FCA)
1937 - Wildlife Restoration Act (WRA)
1958 - Fish and Wildlife Coordination Act (FWCA)
1964 - Wilderness Act (WA)
1965 - Solid Waste Disposal Act (SWDA)
1965 - Water Resources Planning Act (WRPA)
1966 - National Historic Preservation Act (NHPA)
1968 - Wild and Scenic Rivers Act (WSRA)
1969 - National Environmental Policy Act (NEPA)
1970 - Clean Air Act (CAA)
1970 - Occupational Safety and Health Act (OSHA)
1972 - Water Pollution Control Act (WPCA)
1972 - Marine Protection, Research and
        Sanctuaries Act (MPRSA)
1972 - Coastal Zone Management Act (CZMA)
1972 - Home Control Act (HCA)
1972 - Federal Insecticide, Fungicide and
        Rodenticide Act (FIFRA)
1972 - Parks and Waterways Safety Act (PWSA)
1972 - Marine Mammal Protection Act (MMPA)

1973 - Endangered Species Act (ESA)
1974 - Deepwater Port Act (DPA)
1974 - Safe Drinking Water Act (SDWA)
1974 - Energy Supply and Environmental
        Coordination Act (ESECA)
1976 - Toxic Substances Control Act (TSCA)
1976 - Federal Land Policy and Management Act (FLPMA)
1976 - Resource Conservation and Recovery Act (RCRA)
1977 - Clean Air Act Amendments (CAAA)
1977 - Clean Water Act (CWA)
1977 - Surface Mining Control and Reclamation Act
        (SMCRA)
1977 - Soil and Water Resources Conservation Act
        (SWRCA)
1978 - Endangered Species Act Amendments (ESAA)
1978 - Environmental Education Act (EEA)
1980 - Comprehensive Environmental Response
        Compensation and Liability Act (CERCLA)
1982 - Nuclear Waste Policy Act (NWPA)
1984 - Resource Conservation and Recovery Act
        Amendments (RCRAA)
1984 - Environmental Programs and Assistance Act (EPAA)
1986 - Safe Drinking Water Act Amendments (SDWAA)
1986 - Superfund Amendments and Reorganization Act
        (SARA)

33. Khrushchev (1970), 385.
34. *Newsweek,* 8/26/68, 59.
35. *New York Times,* 12/25/68, 1.
36. *New York Times,* 2/6/69.
37. *Baltimore Sun,* 2/12/68.
38. Collier (1996), 169-191; Hollander, 3-80.
39. Hollander, 49-78, 443-468.

## CHAPTER FOURTEEN

1. Cunningham, 187.
2. *U.S. News & World Report,* 11/11/91, 16
3. Newhall, 525, 583, 1090.
4. Newhall, 642, 645.

5. http://www.hq.nasa.gov/osf/1997/com.html
6. *Space News,* 5/18-24/98, 3
7. *Space News,* 2/2-8/98, 11; 3/23-29/98, 16; 5/25-31/98, 6
8. *Space News,* 3/23/29/98, 16
9. *Space News,* 3/23-29/98, 1
10. *New York Times,* 2/8/94, D1; 9/9/94, A7; 9/11/94, IV-2; 11/28/94, D3; 12/18/ 94, V-3; 2/5/95, VI-45; 7/3/95, A4
11. Reagan, 335
12. Harwood, CBS News webpage: ftp://uttm.com/pub/space/STS-86_Archive.txt
13. U.P.I., 9/30/97.
14. http://www.spacedev.com
15. Posted in sci.space.policy 10/9/97.

# EDITORIAL MINUTIAE
## AND GLOSSARY

**Acronyms:** Acronyms have been printed in two ways so that the reader will know how to pronounce them. If the custom is to take the letters and pronounce them as a word (as in NASA), the word is shown with no periods. If the custom was to say the letters, one at a time (as in F.A.A.), periods are inserted.

**Names:** I have generally used the names commonly employed at the time. For example, the Johnson Space Center in Houston was still called the Manned Spacecraft Center in 1968, and so that is the name used in this history. The same applies to Cape Canaveral, which from 1963 to 1973 was renamed Cape Kennedy in memory of John Kennedy.

**Time:** For all events relating to the flight of Apollo 8 I have adopted Central Standard Time, or Houston time. This was the schedule that the astronauts, their families, and mission control lived by and how they experienced the mission. It is also a rough approximation of the day and night schedule for the rest of the country.

**Quotes:** There is no made-up dialogue in this book. In all cases I have either used the exact words as told to me by witnesses or have quoted directly from transcripts. Any changes in dialogue for grammatical reasons is indicated by the use of brackets or ellipses. Similarly, my descriptions of certain individuals' personal thoughts is based entirely on what they themselves remembered thinking.

**Distance:** the space program used nautical miles. I have converted these to statute miles, the measure used by the American public.

**Saturn 5:** While the roman numeral has traditionally been used, I have used Saturn 5 for clarity.

## GLOSSARY OF UNUSUAL TERMS

T.L.I. (Trans-Lunar Injection): The engine burn that lifted the spacecraft out of earth orbit and on its way to the moon. Took place on Saturday morning at 9:40 AM (C.S.T.).

L.O.I. (Lunar Orbit Insertion): The engine burn that put the spacecraft into lunar orbit. Took place Tuesday morning at 4 AM (C.S.T.).

T.E.I. (Trans-Earth Injection): The engine burn that pushed the spacecraft out of lunar orbit and sent it back to the earth. Took place on Thursday morning, ten minutes after midnight.

S.P.S. (Service Propulsion System): The main engine in the service module, used to place the spacecraft in lunar orbit at L.O.I. as well as push it out of lunar orbit at T.E.I. This rocket engine was the only means for the astronauts to leave lunar orbit, and if it failed the men would not be able to return to earth.

Capcom: Shorthand term for CAPsule COMmunications, the individual assigned to handle all ground-to-capsule communications with the astronauts. On Apollo 8 the three capcoms were astronauts Mike Collins, Jerry Carr, and Ken Mattingly.

Command module: The cone-shaped capsule which contained the astronauts' living quarters. This was the only part of the entire rocket that returned to earth.

Service module: Attached to the command module, this cylindrical unit contained most of the spacecraft's tanks and control systems, including the S.P.S. engine. It was abandoned just prior to earth reentry and allowed to burn up in the earth's atmosphere.

I.M.U. (inertial measuring unit): Part of the spacecraft's guidance and navigational system, this device used gimbals and gyroscopes to track the spacecraft's orientation relative to the earth and solar system. Each time the spacecraft's orientation shifted, the I.M.U. recorded the change and indicated this on the eight ball, a specially designed indicator on the instrument panel.

S4B: This was the acronym for the Saturn 5's third stage rocket, used to put the spacecraft into earth orbit during launch, and then fired at T.L.I. to send Apollo 8 on its journey towards the moon.

# BIBLIOGRAPHY

The bulk of my information comes from flight transcripts as well as personal interviews with the astronauts, their wives, and their children and friends. All unfootnoted quotes come from these sources.

For historical background as well as the technical flight details, a number of books deserve special mention. For background on the Cold War, Nikita Khrushchev's memoirs — *Khrushchev Remembers, The Last Testament,* and *The Glasnost Tapes* — were enormously helpful. Also helpful were Michael Bourdeaux's books on religious persecution in the Soviet Union, *Religious Ferment in Russia, Protestant Opposition to Soviet Religious Policy* and *Patriarch and Prophets: Persecution of the Russian Orthodox Church Today.* Norman Gelb's *The Berlin Wall: Kennedy, Khrushchev, and a Showdown in the Heart of Europe* and Pierre Galante's *The Berlin Wall* offered the most information on the tragic history of Berlin during the Cold War.

To get a general perspective of the 1960s and 1968 in particular I found Todd Gitlin's *The Sixties: Years of Hope, Days of Rage* and Peter Collier and David Horowitz's *Destructive Generation, Second Thoughts about the Sixties* useful, especially because the writers were participants in the protests and came to opposite conclusions about their merits. For the Columbia student riots, Roger Kahn's *The Battle for Morningside Heights, Why Students Rebel* is the most objective and thorough description of the event and its participants. I also found the S.D.S.'s newsletter, *New Left Notes*, especially informative for clarifying its point of view.

Only three books had previously described at length the flight of Apollo 8, and all three were crucial reference works: Andrew Chaikin's *A Man on the Moon*, Frank Borman's *Countdown*, and Jim Lovell's *Lost Moon*. Since Chaikin's book told the story of Apollo 8 mostly from Bill Anders's point of view, these books provided me with three different perspectives on the flight. Technical information was best found in David Baker's mammoth *The History of Manned Spaceflight*. *Apollo, the Race to the Moon* by Charles Murray and Catherine Bly Cox tells the story of mission control, the people who made NASA work while still trapped on earth. Jim Oberg's classic work, *Red Star in Orbit*, provides a thorough, though somewhat outdated, view of the Soviet space program.

Above all, the wealth of information provided to me by the NASA History Office in Washington, D.C. cannot be underestimated. Without the help of archivists Mark Kahn and Colin Fries in particular, my job in writing this book would have been much more difficult. I also must thank Meg Hacker at the National Archives in Ft. Worth, Texas for providing me the transcripts for the Gemini 6 and 7 missions, and Robert N. Tice at the Goddard Spaceflight Center, without whose help I would never have been able to discover who took the first earthrise picture.

INTERVIEWS:

THE ASTRONAUTS AND THEIR FAMILIES:

Alan Anders, 12/29/97; Bill Anders, 12/29/97, 2/13/98, 5/9/98; Valerie Anders, 12/30/97; Barbara Borman 1/22/98, 1/25/98; Ed Borman 1/23/98; Frank Borman, 12/4/97, 2/3/98, 3/30/98; Fred Borman, 1/22/98; Susan Borman, 12/4/97, 2/3/98, 5/10/98; Jay Lovell, 2/5/98; Jim Lovell, 12/1/97, 2/2/98, 5/5/98; Marilyn Lovell, 12/2/97, 12/17/97, 2/4/98.

OTHER ASTRONAUTS, FRIENDS, AND CO-WORKERS:

Si Bourgin 10/21/97, 11/20/97, 1/20/98; Jerry Carr, 1/2/98; Mike Collins 1/3/98, 1/20/98; Jim Elkins, 12/12/97; Margaret Elkins, 12/12/97; Dick Gillen 1/25/98; Winnie Gillen 1/25/98; Barbara Gregory 1/17/98; Fred Gregory 12/22/97, 1/17/98,

2/4/98; Adeline Hammack 1/21/98; Jerry Hammack 1/21/98; Dale Klein, 12/8/97; Joe Laitin 10/18/97, 10/19/97, 1/20/98; T.K. Mattingly 1/14/98; Leno Pedrotti 1/26/98; Julian Scheer, 10/23/97, 1/20/98, 1/21/98; Robert Springer, 12/8/97.


OTHER SOURCES:


Adelson, Alan. *SDS.* New York: Charles Scribner's Sons, 1972.

Aldrin, Buzz, and Malcom McConnell. *Men From Earth.* New York: Bantam Books, 1989.

Anderson, John. *Religion, State and Politics in the Soviet Union and Successor States.* Cambridge: Cambridge University Press, 1994.

Anderson, Virginia DeJohn. "Religion, the Common Thread of Motivation," in *Major Problems in American Colonial History,* ed. Karen Ordahl Kupperman. Lexington, Mass.: D.C. Heath and Company, 1993.

Armbrister, Trevor. *A Matter of Accountability, the True Story of the Pueblo Affair.* New York: Coward-McCann, 1970.

Aroneanu, Eugene, ed. *Inside the Concentration Camps, Eyewitness Accounts of the Life in Hitler's Death Camps.* London: Praeger, 1996.

Aronowitz, Stanley. *The Death and Rebirth of American Radicalism.* New York: Routledge, 1996.

*Atlas of the Unknown Face of the Moon.* Moscow: Academy of Sciences, U.S.S.R., 1960.

Ausland, John. *Kennedy, Khrushchev, and the Berlin-Cuba Crisis, 1961-1964.* Oslo: Scandinavian University Press, 1996.

Baker, David. *The History of Manned Space Flight.* New York: Crown Publishers, 1981.

Baker, Michael A., Bradley R. Brewer, Raymond DeBuse, Sally T. Hillsman, Murray Milner, and David V. Soeiro. *Police on Campus, the Mass Police Action at Columbia University, Spring, 1968.* New York: New York Civil Liberties Union, 1969.

Balzhiser, Richard E. "Meeting the Near-Term Challenge for Power Plants" in *Technology and Environment,* pp. 95-113. Washington, D.C.: National Academy Press, 1989.

Barghoorn, Frederick C. *The Soviet Cultural Offensive, the Role of Cultural Diplomacy in Soviet Foreign Policy.* Princeton: Princeton University Press, 1960.

Beschloss, Michael R. *The Crisis Years: Kennedy and Khrushchev, 1960-1963.* New York: Edward Burlingame Books, 1991.

Borman, Frank, with Robert J. Serling. *Countdown, An Autobiography.* New York: Silver Arrow Books, 1988.

Bourdeaux, Michael, *Religious Ferment in Russia, Protestant Opposition to Soviet Religious Policy.* London: MacMillian, 1968.

_____. *Patriarch and Prophets, Persecution of the Russian Orthodox Church Today.* New York: Praeger Publishers, 1970.

Breuer, William B. *Race to the Moon, America's Duel with the Soviets.* Connecticut: Praeger Publishers, 1993.

Brumberg, Abraham, ed. *Russia Under Khrushchev, An Anthology from "Problems of Communism".* New York: Frederick A. Praeger, 1962.

Burlatsky, Fedor. *Khrushchev and the First Russian Spring.* London: Weidenfeld and Nicolson, 1991.

Camp, Glen D. Jr., ed. *Berlin in the East-West Struggle, 1958-1961.* New York: Facts on File, 1971.

Carson, Clayborne. *In Struggle: SNCC and the Black Awakening of the 1960s.* Cambridge: Harvard University Press, 1995.

Cate, Curtis. *The Ides of August, the Berlin Wall Crisis, 1961.* New York: M. Evans & Co., 1978.

Chaikin, Andrew. *A Man on the Moon, The Voyages of the Apollo Astronauts.* New York: Penguin Books, 1994.

Churchill, Winston S. *Closing the Ring.* Boston: Houghton Mifflin, 1951.

Clark, Philip. *The Soviet Manned Space Program.* London: Salamander Books, 1988.

Clarke, Arthur C. *Prelude to Space.* New York: Lancer Books, 1970.

Collier, Peter and David Horowitz. *Destructive Generation, Second Thoughts about the Sixties.* New York: Free Press, 1996.

Collier, Richard. *Bridge Across the Sky, the Berlin Blockade and Airlift: 1948-1949.* New York: McGraw-Hill, 1978.

Collins, Michael. *Liftoff: The Story of America's Adventure in Space.* New York: Grove Press, 1988.

_____. *Carrying the Fire, an Astronaut's Journeys*. New York: Farrar, Straus, and Giroux, 1989.

Compton, William David. *Where No Man Has Gone Before: A History of Apollo Lunar Exploration Missions*. Washington, D.C.: NASA, 1989.

Corley, Felix. *Religion in the Soviet Union: An Archival Reader*. New York: New York University Press, 1996.

Cox, Archibald, et. al. *Crisis at Columbia, Report of the Fact-Finding Commission Appointed to Investigate the Disturbances at Columbia University in April and May 1968*. New York: Vintage Books, 1968.

Cunningham, Walter. *The All-American Boys*. New York: MacMillan Publishing Co., 1977.

Daniloff, Nicholas. *The Kremlin and the Cosmos*. New York: Alfred A Knopf, 1972.

Davison, W. Phillips Davison. *The Berlin Blockade: A Study in Cold War Politics*. Princeton: Princeton University Press, 1958.

Dawisha, Karen. *The Kremlin and the Prague Spring*. Berkeley: University of California Press, 1984.

Dougan, Clark, and Stephen Weiss. *Nineteen Sixty-Eight*. Boston: Boston Publishing Co., 1983.

Dubcek, Alexander, with Andras Sugar. *Dubcek Speaks*. London: I.B. Tauris & Co., 1990.

Dulles, Eleanor Lansing. *Berlin, the Wall is Not Forever*. Chapel Hill: University of North Carolina Press, 1967.

Ellis, Jane. *The Russian Orthodox Church, a Contemporary History*. Bloomington: Indiana University Press, 1986.

Fletcher, William C. *The Russian Orthodox Church Underground, 1917-1970*. London: Oxford University Press, 1971.

Foerster, Norman. *American Poetry and Prose*. Boston: Houghton Mifflin, 1947.

Foner, Philip S., ed. *The Black Panthers Speak*. New York: Da Capo Press, 1995.

Gagarin, Yuri. *Road to the Stars: Notes by Soviet Cosmonaut No. 1*. Moscow: Foreign Languages Publishing House, undated.

Galante, Pierre, with Jack Miller. *The Berlin Wall*. Garden City, N.Y.: Doubleday, 1965.

Gallery, Daniel V. *The Pueblo Incident*. Garden City, N.Y.: Doubleday & Co., 1970.

Gelb, Norman. *The Berlin Wall, Kennedy, Khrushchev, and a Showdown in the Heart of Europe.* New York: Times Books, 1986.

Gitlin, Todd. *The Sixties: Years of Hope, Days of Rage.* New York: Bantam Books, 1987.

Guiley, Rosemary Ellen. *Moonscapes: a Celebration of Lunar Astronomy, Magic, Legend, and Lore.* New York: Prentice Hall, 1991.

Gun, Nerin E. *The Day of the Americans.* New York: Fleet Publishing Company, 1966.

Gunther, John. *Inside Russia Today.* New York: Harper and Brothers, 1958.

Hall, Eldon C. *Journey to the Moon: The History of the Apollo Guidance Computer.* Reston, Va.: American Institute of Aeronautics and Astronautics, Inc., 1996.

Hansen, James R. *Spaceflight Revolution NASA Langley Research Center from Sputnik to Apollo.* Washington, D.C.: NASA SP-4308, 1995.

Hayden, Tom. *Rebellion and Repression: Testimony by Tom Hayden before the National Commission on the Causes and Prevention of Violence, and the House Un-American Activities Committee.* New York: World Publishing, 1969.

_____, *Trial.* New York: Holt, Rinehart, and Winston, 1970.

_____, *Reunion.* New York: Random House, 1988.

Hayward, Max and William C. Fletcher, eds. *Religion and the Soviet State: A Dilemma of Power.* London: Pall Mall Press, 1969.

Heath, G. Louis. *Vandals in the Bomb Factory: The History and Literature of the Students for a Democratic Society.* Metuchen, N.J.: Scarecrow Press, 1976.

Herring, George C. *America's Longest War: The United States and Vietnam, 1950-1975.* New York: McGraw-Hill, 1996.

Hollander, Paul. *Anti-Americanism, Irrational and Rational.* New Brunswick, N.J.: Transaction Publishers, 1995,

Hoyle, Fred. *The Nature of the Universe.* New York: Harper and Brothers, 1960.

Hyland, William and Richard Wallace Shryock. *The Fall of Khrushchev.* New York: Funk & Wagnalls, 1968.

Jacobson, Gary C. *The Politics of Congressional Elections.* New York: Harper Collins, 1992.

Johnson, Priscilla. *Khrushchev and the Arts, the Politics of Soviet Culture, 1962-1964.* Cambridge: M.I.T. Press, 1965.

Kahn, Roger. *The Battle for Morningside Heights, Why Students Rebel.* New York: William Morrow and Co., 1970.

Kamanin, Nikolay. *"I Feel Sorry For Our Guys" -- General N. Kamanin's Space Diaries.* Washington, D.C.: NASA Document TT-21658 1993.

Karnow, Stanley. *Vietnam, a History.* New York: Viking Press, 1983.

Karst, George M. *Beasts of the Earth.* New York: Albert Unger, 1942.

Kelman, Steven. *Push Comes to Shove, the Escalation of Student Protest.* Boston: Houghton Mifflin Co., 1970.

Khrushchev, Nikita. *Khrushchev Remembers.* Boston: Little, Brown and Co., 1970.

_____. *Khrushchev Remembers, The Last Testament.* Boston: Little, Brown and Co., 1974.

_____. *Khrushchev Remembers, The Glasnost Tapes.* Boston: Little, Brown and Co., 1990.

Kousoulas, Dimitrios G. *The Price of Freedom, Greece in World Affairs, 1939-1953.*

Lambright, W. Henry. *Powering Apollo, James E. Webb of NASA.* Baltimore: John Hopkins University Press, 1995.

Lay, Beirne, Jr. *Earthbound Astronauts, the Builders of Apollo-Saturn.* Englewood Cliffs, N.J.: Prentice Hall, Inc., 1971.

Lebedev, L., B. Lyk'yanov, and A. Romanov. *Sons of the Blue Planet.* Moscow: Political Literature Press, 1971.

Lewis, H.A.G. *The Times Atlas of the Moon.* London: Times Newspapers, 1969.

Logsdon, John M. *The Decision to Go to the Moon, Project Apollo and the National Interest.* Cambridge: MIT Press, 1970.

Lovell, Jim. *The Development of the Liquid-Fuel Rocket.* Thesis, United States Naval Academy, 1952.

Lovell, Jim, and Jeffrey Kluger. *Lost Moon, The Perilous Voyage of Apollo 13.* Boston: Houghton Mifflin Co., 1994.

Medvedev, Roy. *Khrushchev.* New York: Anchor Press/Doubleday, 1983.

Medvedev, Roy and Zhores Medvedev. *Khrushchev, the Years in Power.* New York: Columbia University Press, 1976.

Melosh, H. J. *Impact Cratering, a Geologic Process.* New York: Oxford University Press, 1989.

Morgan, Edmund, ed. *Puritan Political Ideas, 1558-1794.* Indianapolis: Bobbs-Merrill, 1965.

Morris, Eric. *Blockade Berlin & the Cold War.* London,:Hamish Hamilton, 1973.

Moss, George Donelson. *Vietnam: An American Ordeal.* Englewood Cliffs, N.J.:
    Prentice Hall, 1994.

Murphy, Edward R., Jr., with Curt Gentry. *Second in Command: Te Uncensored
    Account of the Capture of the Spy Ship Pueblo.* New York: Holt, Rinehart
    and Winston, 1971.

Murray, Charles, and Catherine Bly Cox. *Apollo, the Race to the Moon.* New York:
    Simon and Schuster,

Musgrove, Robert G. *Lunar Photographs from Apollo 8, 10, and 11, SP-246.*
    Washington, D.C.: NASA, 1971.

NASA. *Analysis of Apollo 8s Photography and Visual Observations, SP-201.* Washing-
    ton, D.C.: NASA, 1969.

*New Left Notes,* newsletter for "Students for a Democratic Society," 1966-1971.

Newhall, Christopher G., and Raymundo S. Punongbayan. *Fire and Mud, Eruptions
    and Lahars of Mount Pinatubo, Philippines.* Seattle: University of
    Washington Press, 1996.

Newton, Michael. *Bitter Grain: Huey Newton and the Black Panther Party.* Los
    Angeles: Holloway House, 1991.

Nixon, Richard. *The Memoirs of Richard Nixon.* New York: Grossett and Dunlap,
    1978.

———. *Six Crises.* New York: Simon and Schuster, 1990.

O'Ballance, Edgar. *The Greek Civil War, 1944-1949.* New York: Frederick A. Praeger,
    1966.

Oberg, James E. *Red Star in Orbit.* New York: Random House, 1981.

Pearson, Hugh. *The Shadow of the Panther: Huey Newton and the Price of Black Power
    in America.* New York: Addison-Wesley Publishing Company, 1994.

Pehe, Jiri, ed. *The Prague Spring: a Mixed Legacy.* New York: Freedom House, 1988.

Phillips, Lt. Gen. Sam C. "A Most Fantastic Voyage" in *National Geographic,* May
    1969, 594-631.

Pistrak, Lazar. *The Grand Tactician: Khrushchev's Rise to Power.* New York: Praeger,
    1961.

Reagan, Ronald, *A Time for Choosing: The Speeches of Ronald Reagan, 1961-1982.*
    Chicago: Regnery Gateway, Inc., 1983.

Schick, Jack M. *The Berlin Crisis, 1958-1962.* Philadelphia: University of Pennsylva-
    nia Press, 1971.

*Science for the People, Bi-monthly Publication of Scientists and Engineers for Social and Political Action,* 1-7 (1970-1975).

Seale, Bobby. *Seize the Time: the Story of the Black Panther Party and Huey P. Newton.* Baltimore: Black Classic Press, 1991.

Shadrake, Alan. *The Yellow Pimpernels: Escape Stories of the Berlin Wall.* London: Robert Hale and Co., 1974.

Shepard, Alan, and Deke Slayton with Jay Barbree and Howard Benedict. *Moon Shot: The Inside Story of America's Race to the Moon.* Atlanta: Turner Publishing Inc., 1994.

Shlapentokh, Vladimir. *Soviet Intellectuals and Political Power, the Post-Stalin Era.* Princeton: Princeton University Press, 1990.

Sinclair, Barbara. *The Transformation of the U.S. Senate.* Baltimore: John Hopkins University Press, 1989.

Skolnick, Jerome H. *The Politics of Protest, report prepared for the National Commission on the Causes and Prevention of Violence.* New York: Ballantine Books, 1969.

Slayton, Donald K. "Deke", with Michael Cassutt. *Deke! U.S. Manned Space: From Mercury to the Shuttle.* New York: Tom Doherty Associates, 1994.

Struve, Nikita. *Christians in Contemporary Russia,* London, Harvill Press, 1963.

Titov, Gherman. *Gherman Titov, First Man to Spend a Day in Space.* New York: Crosscurrents Press, 1962.

Tocqueville, Alexis de. *Democracy in America.* New York: Harper & Row, 1969.

Unger, Irwin and Debi. *The Movement, History of the American New Left, 1959-1972.* New York: University Press of America, 1974.

U.S. Department of State, Office of Public Services. *Background: Berlin, 1961.* Washington, D.C.: Government Printing Office, 1961.

Verne, Jules. *The Omnibus Jules Verne.* New York: J.B. Lippincott Co., undated.

Vladimirov, Leonid. *The Russian Space Bluff.* London: Tom Stacey, Ltd., 1971.

Vlavianos, Haris. *Greece, 1941-49: From Resistance to Civil War, the Strategy of the Greek Communist Party.* New York: St. Martin's Press, 1992.

Von Braun, Wernher. *Space Frontier.* New York: Holt, Rinehart and Winston, 1971.

Von Braun, Wernher, and Frederick I. Ordway III. *Space Travel, an Update of History of Rocketry and Space Travel.* New York: Harper and Row, 1985.

Whitney, Thomas P. ed. *Khrushchev Speaks, Selected Speeches, Articles, and Press Conferences, 1949-1961.* Ann Arbor: University of Michigan Press, 1963.

Wilford, John Noble. *We Reach the Moon.* New York: W.W. Norton, 1969.

Williams, Kieran. *The Prague Spring and Its Aftermath, Czechoslovak Politics, 1968-1970.* Cambridge: Cambridge University Press, 1997.

Wolfe, Tom. *The Right Stuff.* New York: Bantam Books, 1980.

# INDEX